Mastering WooCommerce

Build, customize, and launch a complete e-commerce website
with WooCommerce from scratch

Patrick Rauland

Mastering WooCommerce

Group Product Manager: Aaron Tanna
Publishing Product Manager: Puneet Kaur
Senior Editor: Esha Banerjee
Technical Editors: Jubit Pincy and Rajdeep Chakraborty
Copy Editor: Safis Editing
Project Coordinator: Deeksha Thakkar
Indexer: Tejal Soni
Production Designer: Ponraj Dhandapani
Dev-Rel Marketing Coordinators: Deepak Kumar and Mayank Singh

First published: March 2019

Second edition: April 2024

Production reference: 1250424

Published by Packt Publishing Ltd.
Grosvenor House
11 St Paul's Square
Birmingham
B3 1RB, UK.

ISBN 978-1-83508-528-8

www.packtpub.com

Contributors

About the author

Passionate about WooCommerce, **Patrick Rauland** has used it as a customer, worked for WooCommerce support, developed core functionality in WooCommerce itself, led three releases, and helped plan their yearly conference (WooConf). He now provides assistance to people by writing his blog, creating courses for LinkedIn Learning, writing books like this, and consulting on large WooCommerce websites. Patrick is also the co-founder of WooSesh, an online conference for WooCommerce developers and store owners. Patrick lives in Denver, Colorado, where you can probably find him at a local coffee shop; typing away on his computer.

Patrick can be reached through his website: `https://speakinginbytes.com/`.

I would like to acknowledge the incredible and generous WordPress community. I am constantly learning and evolving as a developer because of this community. There's a sense of "we're all in this together" that bonds the community and makes the free flow of information possible and beneficial to all.

I would also like to thank team at Packt Publishing: without their talent and dedication, this book would not be what it is. In particular, I would like to thank Esha Banerjee and Deeksha Thakkar for having faith in this book from the beginning. Adapting their insightful comments raised the quality of this book, and I am grateful for all the time and effort they put into this book.

I would also like to thank the technical reviewer, Nitin Prakash, and the technical editors, for their thorough attention to the programming aspect of this book. Their detailed labels and understanding of target audiences, along with their invaluable comments, greatly improved the clarity of this book.

A special thanks to all of my students for all of their questions. You constantly challenge me to learn and grow which I will always appreciate.

Finally, a special thank you to my family and especially my wife, Ren Rauland, who supported me through this process of spending many nights and weekends writing this book.

About the reviewer

Nitin Prakash, an experienced WordPress plugin and WooCommerce extension developer with over seven years in the field, specializes in creating custom solutions that meet clients' online needs. Proficient in WordPress coding standards, API calls, GIT, SVN, and more, Nitin delivers high-quality development work that aligns with the latest industry best practices. With plugins boasting 10,000+ active installs and 95,000+ downloads on `WordPress.org`, Nitin's track record speaks volumes about his ability to surpass client expectations.

Nitin can be reached at `nitin247@outlook.com`.

Huge thanks to my family, colleagues, and the WordPress community for their support and guidance in my journey. Your contributions have been invaluable.

Additionally, I am grateful for the trust and opportunities provided by my clients, whose diverse needs have challenged me to continually evolve and innovate. Their feedback and collaboration have been essential in refining my skills and delivering solutions that exceed expectations.

Nitin Prakash

Table of Contents

Preface   xiii

Part 1: Exploring the essentials of an ECommerce Store

1

Installing WordPress and WooCommerce   3

Technical requirements   3   Disable marketplace suggestions   9
                                                         Making your own custom plugin   9
New features in WooCommerce   4
WooCommerce Admin   4   Installing WooCommerce   10
WooCommerce Payments   4   Setting your store address and store details 12
WooCommerce blocks   4   Payment settings   14
HPOS   5   How many payment gateways?   15
                                                         Installing Stripe   16
Importance of test sites   5   Add tax rates   17
Migrating the files but not the database   6   Get more sales   20
Testing with a publicly accessible URL   7   Personalizing your store   24
Creating an ad-free experience   7   Summary   27
Jetpack without promotions   8

2

All About Configuring Products   29

Simple products   29   Shipping   33
Product data fields   30   Understanding taxonomies   34
Optional fields   31   Adding images   35
Inventory and stock   32   Product description and short description 37

A fully configured simple product   38

Variable products   40

Attributes   41

Variations   43

Editing individual variations   45

Images for variations   48

Multi-attribute variations   48

Troubleshooting variations   49

Digital products   50

Defining digital products   50

Downloadable but not virtual   51

Configuring virtual and downloadable products   51

Large downloadable files   53

Accessing downloads   53

Product bundles   54

Grouped products   54

Product bundles   56

Configurable bundles   58

Product kits   60

Subscriptions   60

Creating a subscription product   61

Adding a recurring payment option to a product   62

Subscription settings   63

Manual versus automatic renewals   64

Subscription switching   65

Synchronization   65

Retrying failed payments   65

Summary   66

3

Organizing Products   67

Technical requirements   68

Categorizing and tagging   68

Mutually exclusive and collectively exhaustive   69

Tagging products   70

Optimizing product archive pages   71

Writing descriptions for product categories   72

Context matters   74

Meta-description for categories   74

A pretty category description   75

URLs   75

Redirects in WordPress   76

Adding product filters to your Shop page   77

Hidden widgets   82

Active product filters   83

Understanding product blocks   84

Single product pages   89

Customizing product blocks   89

Featuring a product   90

The value of blocks   90

Summary   91

4

Attracting Traffic with Search Engine Optimization   93

Technical Requirements   93

Why you should invest in SEO   94

One-off marketing strategies   94

Always-on marketing   95

Keyword research for e-commerce   95

Creating a list of keywords   96

Comparing search volume   97

Optimizing for keywords   98

Configuring breadcrumbs for search engines and users   98

Adding custom PHP code   100

Using a plugin   102

Creating and sharing an XML sitemap   102

Submitting an XML sitemap to Google   104

Keeping an eye on Google Search Console   105

Summary   107

Part 2: Managing an Online Store

5

Managing Sales Through WP Admin   111

Technical requirements   111

Fulfilling orders   112

Exploring new order notifications   112

New order badge in the site admin   113

Browsing orders   113

Viewing shipping information   114

Packing the boxes   116

Printing shipping labels   118

Dropping off packages   120

Marking orders as complete   121

Refunding orders and payments   121

Refund requests   122

Building a refund process   124

Viewing sales data   125

WooCommerce analytics   126

What sells   129

Gross profit   129

Taxes   130

Using third-party reporting platforms   130

Exploring Metorik   131

Pick one   132

Summary   133

6

Syncing Product Data   135

Exporting out of WooCommerce   136

Exploring a CSV file   136

Including content in a CSV file   138

Importing products via CSV   139

Importing a CSV   139

Integrating with an ERP   142

Finding an ERP   142

Configuring Finale Inventory   144

Importing products into ERP   146

Using an ERP   148

Summary   148

7

Configuring In-Store POS Solutions — 149

Technical requirements	150	Syncing data	158
Setting up WooCommerce POS	150	Syncing data in-store and with WooCommerce	161
Accepting credit cards	152	Single database systems	161
Setting up payment for Stripe	152	Mastering synced databases via an API	162
Selecting WooCommerce POS	154	Manually syncing data	162
Setting up Square	154	Summary	163
Connecting with Square	154		
Setting up Square for WooCommerce	156		

8

Using Fulfillment Software — 165

Sending and updating shipping information	166	Signing up for Shippo	171
		Configuring Shippo's setup information	173
Sending shipping data	166	Fulfilling orders with Shippo	174
Sending emails	166	Configuring ShipStation	177
Configuring webhooks	167	Integrating with ShipStation	177
Building a custom integration	169	Fulfilling packages with ShipStation	181
Updating data	169	Printing pick lists	184
Processing a daily email	170	Using the ShipStation app	185
Retrieving order data through a custom integration	170	Summary	187
Configuring Shippo	171		

9

Speeding Up Your Store — 189

Technical requirements	190	Testing changes	194
Monitoring speed and performance	191	Minifying CSS and JavaScript resources	201
Finding a starting point with GTmetrix	191	Setting up Autoptimize	201
Web Vitals	193		

Concatenate files if necessary   204

Optimizing images   **205**

Optimizing images with Jetpack   205

Optimizing images with Imagify   208

Using the bulk updater   210

Caching and e-commerce   **212**

Configuring caching plugins   212

Configuring caching via HTACCESS   213

Page caching   214

Optimizing content above the fold   **214**

Summary   **216**

Part 3: Customizing the Appearance and Functionality of Your Store

10

Setting Up Your Theme   219

Choosing a theme for WooCommerce   **219**

Exploring the Twenty Twenty-Four theme   220

Storefront   225

Exploring Astra   228

Rearranging the product page   **230**

Installing hook visualizers   231

Stop showing hooks   233

Browsing through code for actions   233

Demo – move the product price   234

Adding a product data tab   **236**

Installing a custom tab plugin   236

Adding a custom tab   237

Extensive customizations using child themes   **239**

Summary   **240**

11

Customizing the Product Page   241

Adding social proof (FOMO)   **241**

Setting up FOMO   242

Customizing notifications   250

Removing events   250

Adding a video tab   **253**

Installing a video tab   254

Adding an extra tab   254

Displaying 360-degree images   **258**

Installing WooCommerce 360° Image   258

Adding 360-degree images to products   259

Summary   **260**

12

Building a Landing Page   261

Building a long-form landing page   262
Creating a new page   262
Understanding the structure of a landing page   263
Add content to a landing page   265

Adding e-commerce to a landing page 268
Adding a featured product   268
Adding an add-to-cart button   270

Finding the product ID   270
Adding the button   271
One Page Checkout   272

Measure and test everything   272
An overview of a CRO experiment   273
Setting up scroll maps and heatmaps   274

Summary   279

13

Creating Plugins for WooCommerce   281

Technical requirements   281
Building a basic WooCommerce plugin 282
Creating a plugin   282
Checking whether WooCommerce is active 283

Customizing order statuses   285
Using the WooCommerce example plugin   286

Registering a post status and adding it to WooCommerce   287

Building a settings page with WooCommerce   289
Creating the main integration file   290
Creating the Integration child class   293
Creating a constructor   293
Adding field settings   295

Summary   300

14

Next Steps with WooCommerce   301

Technical requirements   301
Why and how to make your WooCommerce store accessible   302
Inclusivity   302
Legal liability   303
Business benefits   304
Prepare for demographic trends   305

How to make your store accessible   305

Keeping WooCommerce safe and secure 308
Check your hosting before you launch   308
Use an SSL certificate   309
Keep WordPress core and plugins up to date 309
Keep your version of PHP supported   309

Two-factor login for administrators   310

Scan for downtime   312

Staying up to date with WooCommerce and open source software   312

Follow the Developer Blog   312

Annual conferences   313

Contributing to WooCommerce   314

Office hours   318

Summary   319

Index   **321**

Other Books You May Enjoy   **330**

Preface

According to BuiltWith (`https://trends.builtwith.com/websitelist/WooCommerce-Checkout`) there are over 3,000,000 stores running WooCommerce! That is second only to Shopify with 4 million stores.

There's no one reason people choose WooCommerce. Some store owners like the thousands of themes they have access to, some like that they can build on top of their existing WordPress sites, and others really like the open source customizable angle and knowing that they can always customize the code any way they want. If you decide to build a store on WooCommerce, be prepared for a virtually overwhelming amount of choices. Because it's so popular, you'll always have a lot of options in front of you.

Mastering WooCommerce takes you from an empty WordPress site to having a fully functioning store. As the title of the book implies, we will go deep into WooCommerce and show you the basic options as well as some of the more advanced customizations. We'll do so in an orderly way, starting from the very beginning by setting up a test WordPress site and covering fundamental topics that we'll revisit throughout the rest of the book. Each chapter that follows will expand on the basics, allowing for a gentle progression curve that will allow almost any user to follow along. Each chapter will cover a new section of WooCommerce and thus can be seen as an independent unit, letting you tackle each section separately from the others.

We'll first introduce you to the basics of WooCommerce and WordPress, which will help you develop and debug any issue. You will then learn how to create a simple product and optimize it for SEO. We will then look at shipping, taxes, and payment. After that, we will look into integrating with third-party services for fulfillment and reporting. Furthermore, we will also dive into Point of Sale (POS) systems that let you sell in person. Near the end of the book, we'll create a custom plugin that you can use for any customizations you wish to make. To end the book we take a peek at some advanced topics such as keeping your store safe, making an accessible store, and staying up to date with open source software.

Who this book is for

Mastering WooCommerce is made for everyone who builds WooCommerce sites. You could be a developer who builds sites for clients or you could be a store owner who wants to take a DIY approach with your own store.

You should be familiar with the fundamentals of WordPress. That means understanding what plugins do, what a theme does, how to install plugins and themes, how to keep your site up to date, and how to create posts and pages. So, if you can do the basics with WordPress, *Mastering WooCommerce* will show you the rest.

What this book covers

Chapter 1, Installing WordPress and WooCommerce, brings everyone up to speed. If you've never installed WooCommerce, we're going to go through it together – step by step – looking at setting up our store settings, how we're going to accept payments, and how we're going to keep the admin interface clean.

Chapter 2, All About Configuring Products, digs into all of the settings for products. We'll cover when you should use certain product types, how to give your visitors as much information as possible, and some of the premium product types such as WooCommerce Subscriptions.

Chapter 3, Organizing Products, explains how to add categories and tags to your store and looks into when you would want to do so. We will delve into intuitively organizing your products and enabling customers to find your products and checkout in a flash.

Chapter 4, Attracting Traffic with Search Engine Optimization, is all about getting traffic. We're going to look into some of the common ways to bring people to your site, focusing especially on Search Engine Optimization (SEO) and content marketing, which is very popular with WordPress.

Chapter 5, Managing Sales Through WP Admin, helps store owners manage and fulfill sales. Once you get that traffic, you'll have orders and will need to ship your products. There are some subtle features in the WooCommerce admin realm that make this process surprisingly easy.

Chapter 6, Syncing Product Data, illustrates how tricky it is to keep all of your product data in sync. We'll cover a manual process that you can use to update your products and investigate services that do this for you automatically.

Chapter 7, Configuring In-Store POS Solutions, will highlight the different ways in which you can sell products in person. We lay out several POS solutions, along with their benefits and drawbacks, to help you choose the right one for your store or your client's store.

Chapter 8, Using Fulfillment Software, draws together another suite of tools, this time focusing on fulfillment (getting a package to a customer's door). There are built-in options that are great for starting out but at a certain point, you'll want to use a 3rd party service to save you money and time.

Chapter 9, Speeding Up Your Store, highlights several ways in which you can speed up your store and talks about the importance of doing so. If your store is slow, no one will want to check out, so look into these techniques to speed up your store.

Chapter 10, Setting Up Your Theme, will show you my two favorite themes for WooCommerce and how you can set them up to display your products.

Chapter 11, Customizing the Product Page, is all about building that perfect product page. We'll look into 360-degree images, videos, and the process of adding social proof.

Chapter 12, Building a Landing Page, will show you some key principles of landing page design, how you can add e-commerce functionality to the landing page, and then how to measure the effectiveness of the page and how to optimize it.

Chapter 13, Creating Plugins for WooCommerce, is here for the developers or to give a store owner insight into how a developer does their work. We'll write code to modify WooCommerce itself, customize an order status, build a plugin, and integrate a service with WooCommerce.

Chapter 14, Next Steps with WooCommerce, gives you a look ahead into further areas you can specialize in. We'll look into the importance of accessibility and how you can build an accessible store, how you can keep your WooCommerce store safe and secure, and how you can follow the development of WooCommerce, and contribute code to WooCommerce itself.

To get the most out of this book

You will need to have a functioning WordPress site. Ideally, you should know how to create a test or development site, since many of our examples will change the frontend of your site and you don't want your visitors seeing a work-in-progress store.

I highly recommend that you always keep WordPress and WooCommerce in their latest versions, as well as all the plugins and themes. The further you are behind the latest version, the more features won't work. It's also helpful, but not necessary, to have familiarity with HTML, CSS, JavaScript, and PHP.

Software/hardware covered in the book	Operating system requirements
WordPress 6.5	Windows, macOS, or Linux
WooCommerce 8.8	Windows, macOS, or Linux

If you are using the digital version of this book, we advise you to type the code yourself or access the code from the book's GitHub repository (a link is available in the next section). Doing so will help you avoid any potential errors related to the copying and pasting of code.

Download the example code files

You can download the example code files for this book from GitHub at `https://github.com/PacktPublishing/Mastering-WooCommerce-`. If there's an update to the code, it will be updated in the GitHub repository.

We also have other code bundles from our rich catalog of books and videos available at `https://github.com/PacktPublishing/`. Check them out!

Conventions used

There are a number of text conventions used throughout this book.

`Code in text`: Indicates code words in text, database table names, folder names, filenames, file extensions, pathnames, dummy URLs, user input, and Twitter handles. Here is an example: " So, you can do some digging by searching for `{{your industry}} + POS systems`.

A block of code is set as follows:

```
class WC_Example {
    public function __construct(){
        // add more code here
    }
}
// Start running our plugin
$GLOBALS['wc_example'] = new WC_Example();
```

Any command-line input or output is written as follows:

```
$this->form_fields = array(
    'api_key' => array(
        'title'       => __( 'API Key', 'woocommerce-integration-demo'
),
        'type'        => 'text',
        'description' => __( 'Enter with your API Key. You can find
this in "User Profile" drop-down (top right corner) > API Keys.',
'woocommerce-integration-demo' ),
        'desc_tip'    => true,
        'default'     => '',
    ),
);
```

Bold: Indicates a new term, an important word, or words that you see onscreen. For instance, words in menus or dialog boxes appear in **bold**. Here is an example: "In FOMO, you can click on Events and you can see all of the events that FOMO picked up."

> **Tips or important notes**
> Appear like this.

Get in touch

Feedback from our readers is always welcome.

General feedback: If you have questions about any aspect of this book, email us at `customercare@packtpub.com` and mention the book title in the subject of your message.

Errata: Although we have taken every care to ensure the accuracy of our content, mistakes do happen. If you have found a mistake in this book, we would be grateful if you would report this to us. Please visit `www.packtpub.com/support/errata` and fill in the form.

Piracy: If you come across any illegal copies of our works in any form on the internet, we would be grateful if you would provide us with the location address or website name. Please contact us at `copyright@packt.com` with a link to the material.

If you are interested in becoming an author: If there is a topic that you have expertise in and you are interested in either writing or contributing to a book, please visit `authors.packtpub.com`.

Share Your Thoughts

Once you've read *Mastering WooCommerce*, we'd love to hear your thoughts! Scan the QR code below to go straight to the Amazon review page for this book and share your feedback.

`https://packt.link/r/1835085288`

Your review is important to us and the tech community and will help us make sure we're delivering excellent quality content.

Download a free PDF copy of this book

Thanks for purchasing this book!

Do you like to read on the go but are unable to carry your print books everywhere?

Is your eBook purchase not compatible with the device of your choice?

Don't worry, now with every Packt book you get a DRM-free PDF version of that book at no cost.

Read anywhere, any place, on any device. Search, copy, and paste code from your favorite technical books directly into your application.

The perks don't stop there, you can get exclusive access to discounts, newsletters, and great free content in your inbox daily

Follow these simple steps to get the benefits:

1. Scan the QR code or visit the link below

https://packt.link/free-ebook/9781835085288

2. Submit your proof of purchase
3. That's it! We'll send your free PDF and other benefits to your email directly

Part 1: Exploring the essentials of an ECommerce Store

In *Part 1* we will familiarize ourselves with the essentials of an online store. This section will cover the following chapters:

- *Chapter 1, Installing WordPress and WooCommerce*
- *Chapter 2, All About Configuring Products*
- *Chapter 3, Organizing Products*
- *Chapter 4, Attracting Traffic with Search Engine Optimization*

1

Installing WordPress and WooCommerce

WooCommerce was designed as a WordPress plugin from its conception. Everything that WooCommerce has done is done on top of WordPress. So, while this is a book about mastering WooCommerce, we can't start talking about WooCommerce until we make sure a few basic things are taken care of in your WordPress installation.

We're going to make sure your WordPress site is set up correctly and then install WooCommerce. To do that, we're going to look into the following:

- New features in WooCommerce
- Why and how you should use test sites
- Creating an ad-free admin experience
- How to install WooCommerce
- Configuring settings through the WooCommerce welcome wizard

Once you've done all of the preceding, you'll have WooCommerce installed on a test site and you can start building your online store. Let's first look at why and how we should use test sites with any WordPress installation.

Technical requirements

We'll be installing a few different pieces of software in this chapter:

- WordPress: `https://wordpress.org/`
- WooCommerce: `https://wordpress.org/plugins/woocommerce/`
- Jetpack Without Promotions: `https://wordpress.org/plugins/jetpack/`

- Surbma | WooCommerce Without Marketplace Suggestions: `https://wordpress.org/plugins/surbma-woocommerce-without-marketplace-suggestions/`

- The code files for this chapter can be found in the following GitHub repository: `https://github.com/PacktPublishing/Mastering-WooCommerce-/tree/main/Chapter01`

New features in WooCommerce

The first version of this book came out back in the Spring of 2020. Since then, there have been 30+ major releases of WooCommerce. Let's look at what they've been working on!

WooCommerce Admin

The WooCommerce Admin plugin started with WooComemrce 4.0 and was shipped with WooCommerce 6.5. WooCommerce Admin offers significantly enhanced reporting. It enables filtering and comparisons and provides an overview of your store's performance in terms of sales and revenue.

Additionally, it offers insights into the most popular and highest-grossing products while also enhancing customer management and analytics capabilities.

WooCommerce Payments

WooCommerce also added WooCommerce Payments in 4.1. This feature brings a new dashboard to your site's WordPress administration area for handling payments, refunds, disputes, and deposits.

WooCommerce blocks

With WordPress 5.0 they introduced the new block-based editing experience called Gutenberg. Since then, WordPress software including WooCommerce has been refactored to let users control exactly how they want to lay out their pages with blocks.

The following are just a few of the individual blocks that have been worked on:

- Cart and Checkout blocks

- Single Product Detail block

- Add to Cart Form block

- Blockified Single Product Template

- Dozens of block-based patterns

We're going to cover some of the WooCommerce blocks in this book. Look for *Chapter 10, Setting Up Your Theme,* to see how to use these new blocks.

HPOS

My personal favorite feature is **High-Performance Order Storage** or **HPOS**. Most of the readers of this book won't need this feature immediately; it's more for large stores doing hundreds of orders a day, where the orders can fill up the database quickly, slowing down the whole site.

This HPOS feature builds a specialized Orders table that makes it faster for your store to process new orders and read orders in the WordPress administration area.

Essentially, this feature means that you can start your e-commerce journey on WooCommerce, and can stay on WooCommerce once you've hit your stride and you're making seven figures a year with your e-commerce store.

- WooCommerce was already powerful before these changes, and now it's even better, offering a much-improved admin experience, Block-based functionality to customize your product pages and your theme, and a robust infrastructure update so you can continue using WooCommerce even if your store becomes mega popular.

- If you want to see where WooCommerce is going, there is a public roadmap: `https://developer.woocommerce.com/roadmap/`

Importance of test sites

If you've been a WordPress developer for a while, you're probably familiar with test sites. And while they're important in developing non-e-commerce WordPress sites, they're critical in WooCommerce development. The following screenshot shows what the website development process looks like:

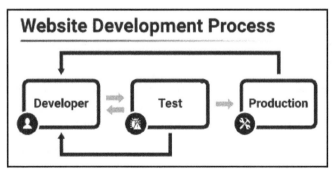

Figure 1.1 – Website development process

In a typical WordPress development project, you'll build custom functionality on your local machine. Then, you'll upload it to a test site where the client usually approves it. Then, you'll move the test site to the live site, replacing data and files.

And this works great for most WordPress projects. But when it comes to e-commerce, there are two problems:

- **You can never replace the live database**: Since an e-commerce site is always on and always accepting new orders and payments, and marking items as shipped, you can't replace the live database with the test database.

- **E-commerce functionality often needs a publicly accessible URL to work properly**: A lot of e-commerce functionality (shipping, payment, and taxes) interacts with third parties, some of which need a publicly accessible URL to return data. So, it's much harder to test your site on a local machine.

Since e-commerce sites have greater demands, we're going to cover some of the things you need to do with a test site:

- Migrate files but not the database

- Test with a publicly accessible URL

With these two extra criteria met, both of which can be done by a good website host, you can easily test and launch your own WooCommerce site. Let's look into migrating files first.

Migrating the files but not the database

With any sort of e-commerce site, it's always on and ready to accept new orders and payments and mark items as shipped. Because of this, if you ever replace a live database with a test database, you could have catastrophic results.

It will often take days or weeks to make a test site, test the changes, and get them approved. In that time frame, there will very often be a new order and if you replace the live database with the test database, you erase all records of that order. That's a bad place to be.

This is why you never want to overwrite a live database. You'll want to work with a host that can let you move your code to your live site and leave the live database intact. Or you may want to have your own processes to quickly move all files from your test site to your live site.

There are a couple of hosts worth mentioning that have powerful infrastructure that helps you to build great WooCommerce sites:

- WP Engine (`https://wpengine.com/`)

- Pantheon (`https://pantheon.io/`)

- Pressable (`https://pressable.com/`)

These hosts will be able to help you to migrate just the files you want without moving the database. If you want to use another host, just make sure it has the infrastructure to migrate files between a test and a live website.

And if you're wondering why I'm mentioning hosts instead of local development software, that's because it's important in e-commerce to develop sites with a publicly accessible URL, as we'll cover in the next section.

Testing with a publicly accessible URL

When you're working on a WooCommerce site, you'll need to test all of the e-commerce functionality, such as getting shipping rates, importing tax rates, and accepting payment.

Unfortunately, some of these third parties use legacy systems to deliver data to your site. And for some of these systems to work, they deliver data to your site via a publicly accessible URL. For example, a shipping company might return data to your store about a custom shipping price with a link similar to this: `yourstore.com/?custom_parameter=foo`

If they can't access your store via a URL, these services might not work. So, if you want to develop a custom theme or plugin that interacts with the cart or checkout, you might have to do that development on a test site instead of a local site on your own computer.

If you are doing a lot of custom development, it still saves time to develop on your local machine, and when you want to test the site, move all of your local files to your test site. But for many e-commerce sites, you can save time by doing all of your development on a test site and skipping the local site.

Now that we know how to develop sites, let's make sure our admin is free from promotions.

Creating an ad-free experience

Both WooCommerce and Jetpack, a plugin we'll install later in this chapter, include promotions. And these promotions make it less clear what's going on. And if you're developing this site for a client, you want to recommend plugins—you don't want your plugin doing that for you.

As an example, in the following screenshot, there's a promotion for premium functionality:

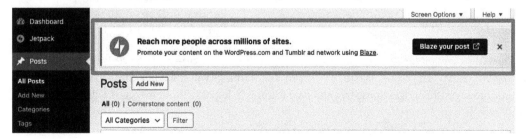

Figure 1.2 – An ad from Jetpack

> **Note**
>
> Throughout this book, I'll include several screenshots. To make sure these are valuable to you, I'm going to make sure they're showing you what I need you to see and I'll try and get rid of the extra content.

To make this book clearer, I'm going to install two plugins that remove these ads, which lets me share more useful screenshots and will give you and your clients a much cleaner user experience.

Let's first install something to prevent promotions from Jetpack.

Jetpack without promotions

One of the plugins you can use to remove all of the ads in Jetpack is **Jetpack Without Promotions**. You can get this plugin from WordPress (`https://wordpress.org`). The following screenshot displays how it looks:

Figure 1.3 – Jetpack Without Promotions on the WordPress.org forum

The actual code for this plugin is tiny. There are only a couple of important lines:

```
add_filter( 'jetpack_just_in_time_msgs', '__return_false', 20 );
add_filter( 'jetpack_show_promotions', '__return_false', 20 );
add_filter( 'jetpack_blaze_enabled', '__return_false' );
```

The first line turns off just-in-time messages (`https://developer.jetpack.com/hooks/jetpack_just_in_time_msgs/`). Errors and warnings will still come through normally. Just-in-time messages are nudges to use free and paid features in Jetpack. Those messages will be turned off.

The second line turns off promotions in the plugin search results, which was added in Jetpack 7.1 (`https://wptavern.com/jetpack-7-1-adds-feature-suggestions-to-plugin-search-results`).

The third line turns off promotions for Blaze, an ad network on Tumblr and WordPress.com (`https://jetpack.com/blog/introducing-blaze-find-new-customers-by-promoting-your-best-content/`).

Disable marketplace suggestions

In WooCommerce 3.6, the WooCommerce team announced Marketplace Suggestions (`https://developer.woocommerce.com/2019/04/03/extension-suggestions-in-3-6/`). These inject recommendations for official WooCommerce extensions into the Orders screen and the Products screen for the store owner. They were adjusted just prior to the release and will likely evolve in the next few versions.

There's a plugin on the WordPress site called **Surbma | WooCommerce Without Marketplace Suggestions**, which disables these promotions.

There's only one important line in the plugin:

```
add_filter( 'woocommerce_allow_marketplace_suggestions', '__return_
false');
```

The code to disable promotions is quite simple: one filter that removes them completely.

Making your own custom plugin

Each of the preceding plugins does one small thing very well. I like to call these utility plugins since they do one thing incredibly well. They don't have a user interface, ads, or premium features—they just work.

You could make your own custom plugin for WooCommerce and include the four lines of code from the preceding two plugins and have the same end result.

If you want to be able to use WooCommerce without ads getting in your way, you'll want to install these plugins or create your own. Talking about installing WooCommerce, let's quickly take a look at how to do that in the next section.

Installing WooCommerce

Let's get started by actually installing WooCommerce on our site. Perform the following steps:

1. Search for WooCommerce under plugins in your admin menu:

Figure 1.4 – WooCommerce in the plugin installer

2. Click **Install Now** followed by **Activate**.

3. Click **WooCommerce** in the admin menu. Or if you're familiar with WooCommerce and want to skip the welcome wizard you can click **Start Selling** on your **Dashboard** screen under **WooCommerce Setup**. See *Figure 1.5*.

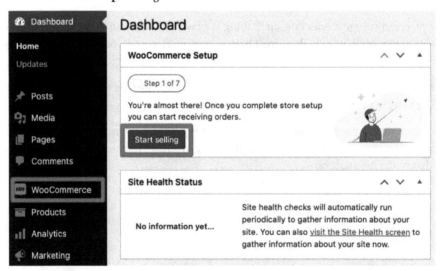

Figure 1.5 – Clicking Start selling will skip the setup wizard

4. Clicking **WooCommerce** will take you to the welcome wizard, which will help you configure all of the settings you'll need to get up and running. Here's what the first step looks like:

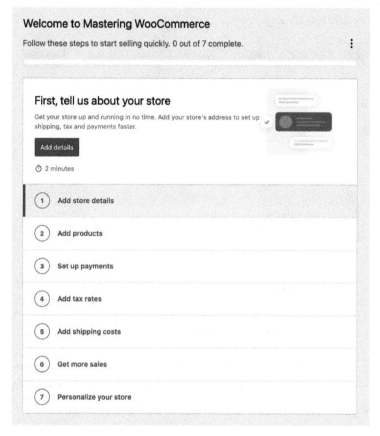

Figure 1.6 – First step of the setup wizard

Now we can configure the store details in WooCommerce. Click **Add details**.

Setting your store address and store details

The first step is adding the address for your store. If you don't have a physical store, add your warehouse address on the **Store Address** settings page:

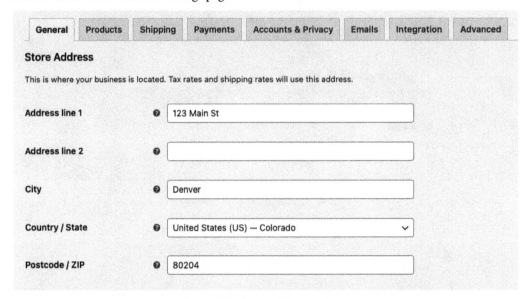

Figure 1.7 – Store address settings

One thing you should know about your address: WooCommerce assumes you have one location for your business. This should be your primary location. WooCommerce uses your location for three features:

Calculating shipping rates via USPS, UPS, FedEx, and other shipping carriers

Importing tax rates

Determining your currency

If you do have multiple locations, you'll have to configure a few extra settings in the shipping and tax settings sections.

When you're done click **Save Changes** at the bottom of the page.

You can then continue the welcome wizard by **WooCommerce | Home** in the admin sidebar or by clicking **Finish setup** in the top right. I'll opt for clicking **Finish setup**:

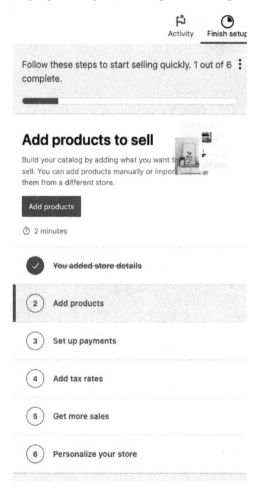

Figure 1.8 – Finish setup will bring you back to the setup wizard

We're going to cover adding a product in *Chapter 2 – All About Configuring Products*. Let's skip that step and click on **Set up payments**.

Payment settings

The next step helps you to turn an online catalog into an online store by being able to take payment. In the settings here, make sure you can actually accept payments:

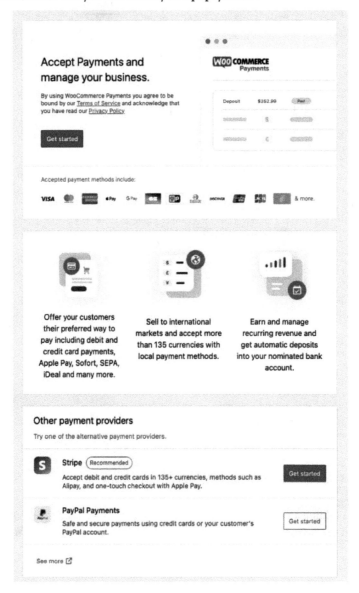

Figure 1.9 – Payment options in WooCommerce

WooCommerce recommends three popular gateways:

- **WooPayments** (`https://woocommerce.com/payments/`)

- **Stripe**, which accepts credit cards and Apple Pay (`https://stripe.com/`)

- **PayPal Payments**, which accepts credit cards, Venmo, and PayPal balances, and also offers a Pay Later option (`https://woocommerce.com/products/woocommerce-paypal-payments/`)

Confusingly, Stripe and WooCommerce Payments are both built on Stripe. WooCommerce Payments creates a Stripe Express account for you, which you access through your WooCommerce store. It's more integrated and easier to set up. It also has better support for subscriptions and multi-currency payments.

However, WooCommerce Payments is also a little more limited in other ways. You can't copy your API keys and connect Stripe to your bookkeeping software or something similar. And if you have multiple e-commerce websites, you can't use the same account. You'd have to create multiple accounts, which can create headaches.

There is an official comparison guide on WooCommerce.com: `https://woocommerce.com/document/woopayments/compatibility/woopayments-vs-stripe-plugin-comparison/`.

For tech-savvy developers I recommend Stripe. You lose very little functionality, and while it's slightly harder to set up, it's much more flexible when it comes to evolving your business in whatever way you need.

There are hundreds of payment gateways available on WooCommerce and WordPress, and sometimes, you need a specialized gateway for a specific currency, locale, or custom payment. However, many sites just want to accept credit cards or PayPal, and for these, Stripe, PayPal, and WooCommerce Payments are perfect.

How many payment gateways?

If you've never set up an e-commerce store, it might be confusing how many payment gateways you need. In short, you only need one payment gateway. However, it's possible for that payment gateway to go down, or more likely for a credit card to be declined. In such cases, it's a great idea to have a backup payment gateway such as PayPal.

If you have the time, set up two payment gateways. If you only have time to set up one payment gateway, I would suggest Stripe – I'm a huge fan since it's very easy to use and test.

Installing Stripe

Click **Get started** next to **Stripe**.

That will install the **Stripe** plugin automatically. And then you'll be prompted to connect your **Stripe** account:

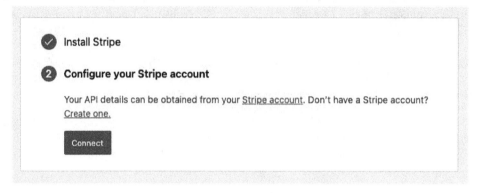

Figure 1.10 – Connect your store to Stripe

Click **Connect** and you'll be prompted to create a free Stripe account or log in to your existing account.

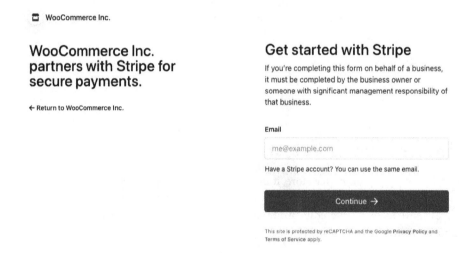

Figure 1.11 – Stripe setup

Once you agree to the terms the Stripe plugin will be installed and configured. And that means our store is ready to accept payment. That's significant progress.

Add tax rates

The fourth step in the welcome wizard is to add tax rates. Click **Add tax rates**.

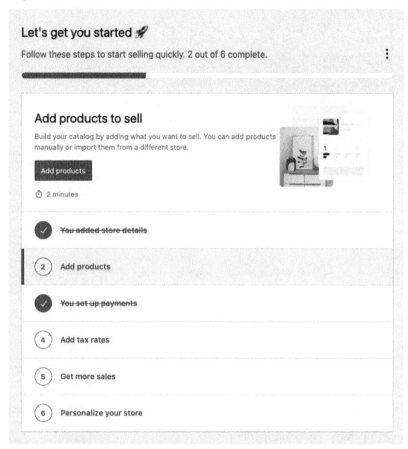

Figure 1.12 – Add taxes in the fourth step of the setup wizard

WooCommerce will give you four options:

- **WooCommerce Tax** (free)
- **Avalara** (third-party subscription required)
- **Set up taxes manually** (free)
- **I don't charge sales tax**

Here are the options you'll see:

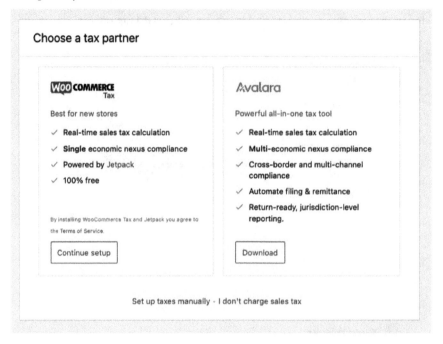

Figure 1.13 – Tax options in the setup wizard

Almost all products are taxable. Unless you've already verified with a tax expert you don't need to collect taxes, make sure you set up some sort of tax rates.

For most stores, WooCommerce Tax will be fine. It will import tax rates for you based on your zip code.

Click **Continue setup** under **WooCommerce Tax**.

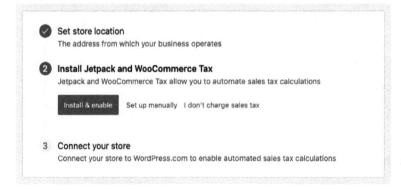

Figure 1.14 – The first step to installing WooCommerce Tax

Then, click **Install & enable**.

It will take a minute to install the Jetpack plugin. Once that's done, click the **Connect** button underneath **(3) Connect your store**.

You should see a success notification after connecting your store to WordPress.com.

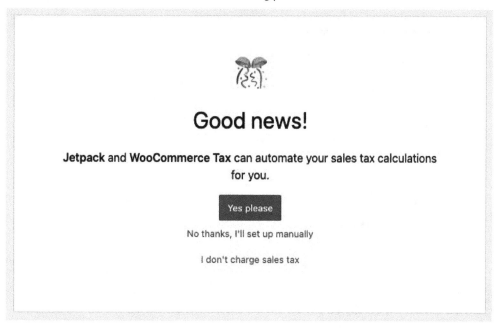

Figure 1.15 – WooCommerce Tax setup complete

Click **Yes please** and you should be done.

> **Note**
>
> If you are having trouble connecting your site to Jetpack, you will need a publicly accessible site. If you're using a local development setup, now is the time to switch to a publicly accessible test site. Additionally, try different browsers. I had trouble connecting to Jetpack with the Brave browser, but it worked just fine in Google Chrome.
>
> If you're a tax aficionado and want to see the tax rates, note that you can't see them until you get a product through the checkout. That's when WooCommerce checks and imports the actual tax rates for your store.

Get more sales

The next section of the welcome wizard is all about recommended add-ons for WooCommerce. Many of these make WooCommerce easier to use and integrate with third-party services. You can see the options here:

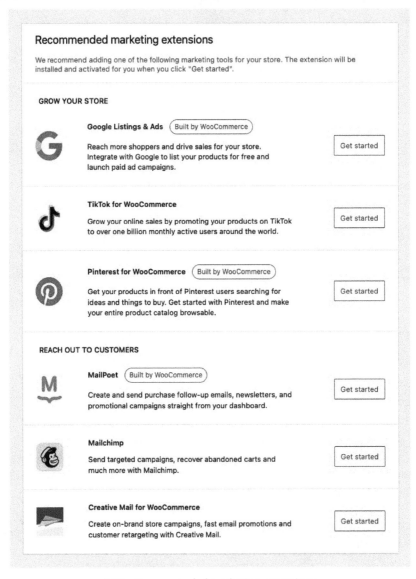

1.16 – Recommended marketing extensions

WooCommerce makes quite a few recommendations:

- **Google Listings & Ads**
- **TikTok for WooCommerce**
- **Pinterest for WooCommerce**
- **MailPoet**
- **Mailchimp**
- **Creative Mail for WooCommerce**

The first three recommendations are for generating awareness. They get your product onto social or ad networks. If you already have a strategy in mind for these platforms, go ahead and install the plugins.

I try to keep my WooCommerce site as minimal as possible. Even if I suspect I may use Pinterest, I'd avoid installing it now and wait till later when I know I need it.

The second group of three extensions is for newsletter services. I recommend you install one of these. If you're going through all the work attracting customers to your store, you should give them a checkbox at checkout so they can choose to hear from you in the future. Many e-commerce merchants generate 10-20% of revenue from their email list. You can see some more email marketing strategies used by seven-figure store owners at `https://www.ecommercefuel.com/ecommerce-email-marketing/`.

If you don't have a newsletter in mind, Mailchimp (`https://mailchimp.com/`) is an excellent place to start. Mailchimp offers a free plan for up to 1,000 subscribers.

> **Note**
>
> If you are building a WooCommerce store for an existing merchant who wants to master email marketing. For this, I would recommend Klaviyo (`https://woocommerce.com/products/klaviyo-for-woocommerce/`). They have powerful features to segment your list based on how recently they purchased and they allow you to build automated flows to remind those users to come back and purchase.

Let's install Mailchimp. Click **Get started** next to Mailchimp. That will install the plugin on your site. Then, click **Manage** to set up the plugin.

Figure 1.17 – Connecting your store to Mailchimp

You should now be able to connect Mailchimp to WooCommerce. I already have an account so I'll use **Connect Account**. You can also **Create Account** if you don't yet have a Mailchimp account.

Log in to your account and give permission to connect to WooCommerce.

> **Note**
>
> I had some trouble connecting my WooCommerce site to Mailchimp. I logged into Mailchimp and went through two-factor security. Then I tried again and it worked. If you get stuck logging in to your Mailchimp account, open a separate window and try again.

Once you do, you should land back on your website.

Figure 1.18 – Add store settings to Mailchimp

Confirm your store settings and click **Next Step**.

On the page that appears, you'll see settings about the newsletter and how users should be imported into Mailchimp.

There is one setting that I recommend all store owners change:

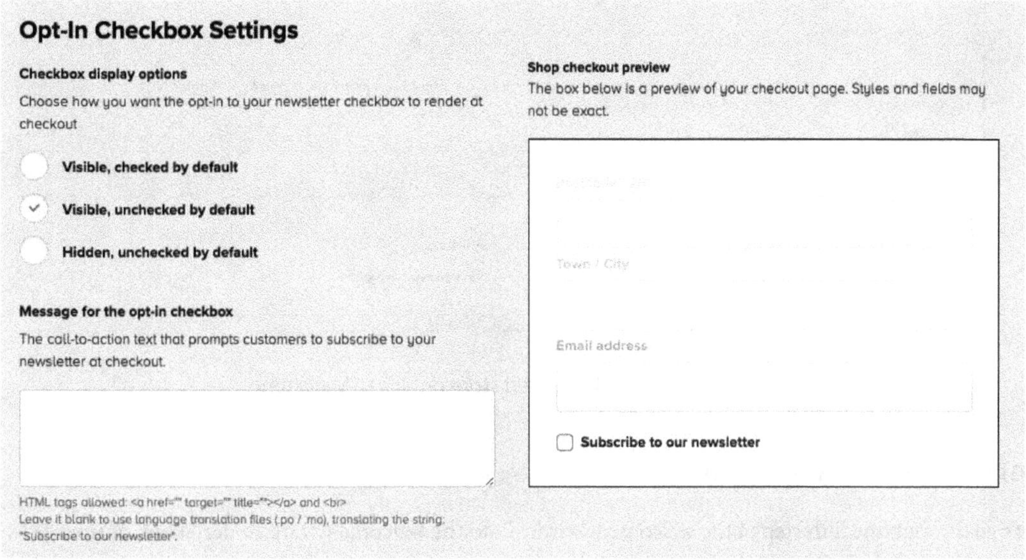

Figure 1.19 – Newsletter opt-in settings

By default, the newsletter checkbox is **Visible, checked by default**. That's called an opt-out because users have to take action to opt-out.

You want to change the setting to **Visible, unchecked by default**. This is an opt-in and you'll get fewer newsletter subscribers but they'll only be users who actually want to hear from you. They'll actually open and reply to your emails, and most importantly, they won't mark your emails as spam, which hurts your whole email list.

Once you've reviewed the other settings, click **Start Sync**.

When you're done, you should see a success page.

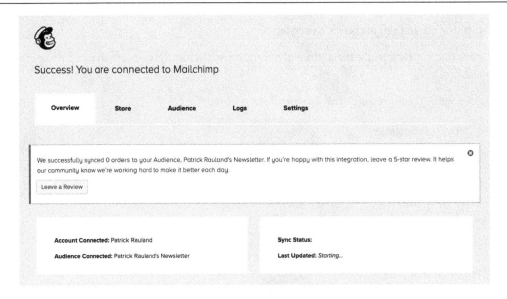

Figure 1.20 – Successfully synced store details to Mailchimp

Personalizing your store

Let's go through the fifth step of the welcome wizard. Go to the welcome wizard under **WooCommerce | Home**, then click on **Personalize your store**.

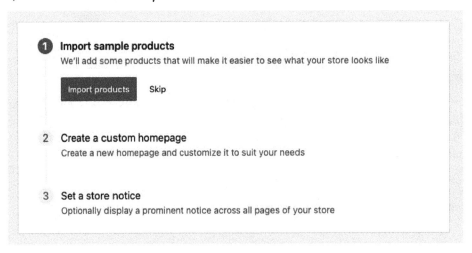

Figure 1.21 – Personalize your store settings

You'll see three steps:

- **Import sample products**
- **Create a custom homepage**
- **Set a store notice**

We're going to cover customizing your theme later in this book. But we'll do the other two steps right now.

Importing sample products will do a good job showing you what your WooCommerce store could look like with a couple dozen products. And you can easily delete them when you want to import all of your products.

Click **Import products**.

It should only take a minute to complete. And then you should get a notice the import is done.

On to the next step! This is the step we're skipping. So, click **Skip**.

Now you'll be prompted to add text under **Set a store notice**. Go ahead and type anything in there. I'm typing this:

```
This is a store notice. Orders over $50 get free shipping.
```

Click **Complete task** and you'll be redirected back to the welcome wizard. You should see a notice that you have completed the setup.

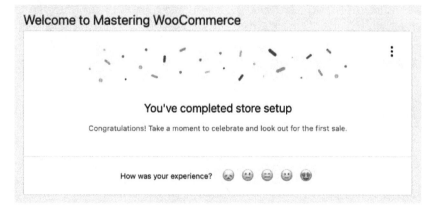

Figure 1.22 – Completing store setup 🎉

Let's also take a look at the frontend of our store. You can get to the frontend of your store from the admin by hovering over your store name in the top left and then clicking **Visit Store**.

Figure 1.23 – Visit Store

And here's what my store looks like:

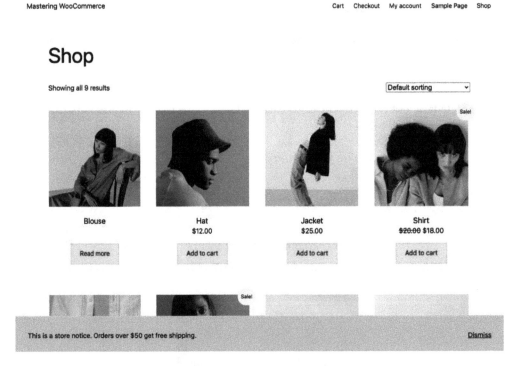

Figure 1.24 – Default store with the Twenty Twenty-Three theme

I'm using the pre-installed theme *Twenty Twenty-Three*. Your store will look very different if you're using a different theme. For now, don't worry about switching themes. You can switch to Twenty Twenty-Three if you want to follow my screenshots but ultimately this is temporary. Later in the book, we'll explore a few different themes and choose one for our store.

Summary

In this chapter, we first went through the steps that we needed to take before installing WooCommerce. We had to ensure that our WordPress site was set up correctly and that we could build everything in a properly set-up test site.

Once we had that sorted, we installed a few plugins to keep our admin interface free of ads, which could distract us or our clients.

We then installed WooCommerce and Jetpack and configured a number of basic settings through the welcome wizard. With these steps, you can set up and test as many WooCommerce sites as you want.

Now that we've done all of that, we're ready to create products in our store. In the next chapter, we will learn to build and configure new products.

2
All About Configuring Products

You can't have an online store without something to sell. Before we can sell something, we have to add products to our store. But before we can even do that, we have to know the different kinds of products we can add to our store.

In this chapter, we're going to look at three types of products that are included in the free WooCommerce plugin. We're also going to look at the *Product bundles* (`https://woocommerce.com/products/product-bundles/`) and *Subscriptions* (`https://woocommerce.com/products/woocommerce-subscriptions/`) that are available through premium extensions on the WooCommerce website. Each of these types of products has unique features, and choosing how you want to display your products in your store is an important decision.

In this chapter, we're going to look into the following topics:

- Simple products
- Variable products
- Digital products
- Product bundles
- Subscriptions

All of these product types build on each other. So, before you jump ahead and look at *Variable products*, make sure you look at the *Simple products* section. By the end of this chapter, you should know what types of products you want to offer in your store, as well as the technical information you need to configure them correctly.

Simple products

The easiest type of product to create is, unsurprisingly, called a **simple product**. It's important to know how to create a simple product because everything else is based on it. So, you'll want to know exactly how to manipulate a simple product. A good example of a simple product would be anything that comes in one format, such as a coffee cup, ring, or backpack. You can add a new product through your WordPress admin under **Products | Add New**.

We're going to look into a few different aspects of simple products:

- **Required and optional fields**: To add useful data to the product page

- **Taxonomies (categories and tags)**: To make it easy for users to browse our store

- **Images**: To instantly communicate what our product looks like

- **Description fields**: To describe our products

Let's explore some of these aspects!

Product data fields

The only required field for a product is **Product name** (also known as the title). If you add the product name to a product and publish it, you'll see an empty page. See *Figure 2.1*:

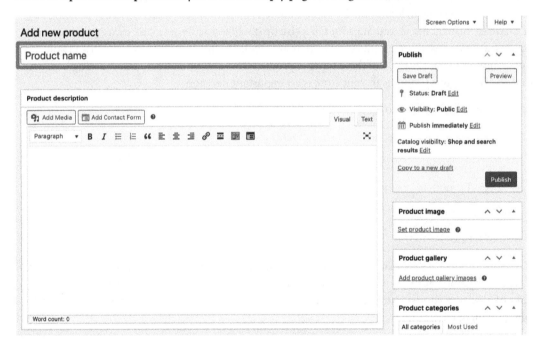

Figure 2.1 – The Add new product page

You can also add a price to the product. It can be 0 or any positive value. As soon as you add a price, users can add it to their cart.

Optional fields

When you take a look at the **Add new product** screen, it might not be clear where you add all of the information about your products. You'll need to scroll down to see the **Product data** panel.

This controls most of the settings for your product, including the following:

- Product type (simple, variable, grouped, and so on)
- Downloadable and/or virtual
- Price
- Shipping dimensions
- Linked products

For now, leave the product type set to **Simple product**. Leave **Virtual** and **Downloadable** unchecked – we'll look at those later in this chapter.

Under the **Product data** panel, set **Regular price** and **Sale price** values:

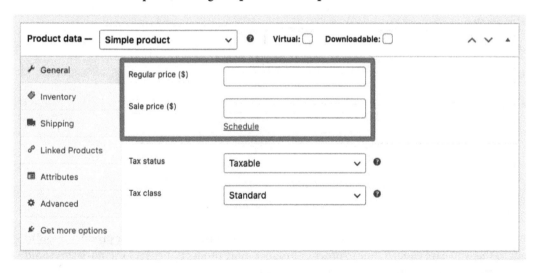

Figure 2.2 – Setting Regular price and Sale price values

After setting a price, you will likely want to look into the inventory and stock settings. These track how much product you have in stock and they ensure that users can only check out while you have items in stock – a very handy tool for a store owner.

Inventory and stock

If you manage stock for your product, you can have WooCommerce automatically track the quantities. Under **Product data | Inventory**, you can enter something for the following:

- Enter a value for **SKU** (stock-keeping unit), which is an ID for a product

- Check **Stock management**

- Once **Stock management** has been checked, you can enter a value for **Quantity**

Here's how I track stock in my store:

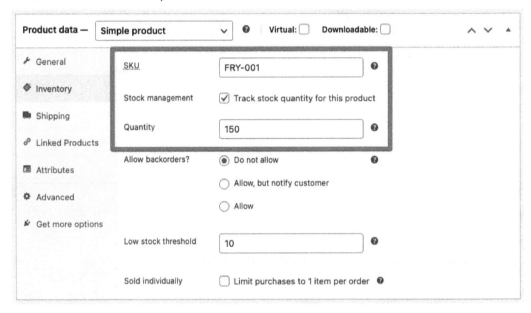

Figure 2.3 – Tracking stock for a product

I'm a big fan of the **Low stock threshold** setting, which will email you when your product is low in stock so that you have enough time to order or produce more products. I've set **Low stock threshold** to 10. However, you can also have a storewide **Low stock threshold setting**, which you can find under **WooCommerce | Settings | Products | Inventory**.

Shipping

The **Shipping** tab in the **Product data** panel is important for two reasons:

- It's what your store uses to calculate live shipping rates

- It's displayed to the customer on the frontend

While images can be helpful to show the size and scale of products, nothing beats having dimensions listed on the product page.

Let's consider an example.

A friend of mine was looking to purchase a new refrigerator and did weeks of research. They found a great model and almost pressed the purchase button when they realized they had no idea of the size and couldn't guarantee it would fit in their kitchen. They ended up purchasing a refrigerator from a different company and the original company lost out on a $1,000+ sale because they didn't list their dimensions.

So, don't forget to list the dimensions!

There are two sets of shipping fields you'll want to fill in:

- **Weight**

- **Dimensions (Length**, **Width**, and **Height)**

So, let's add shipping settings:

Figure 2.4 – Adding shipping settings to the Shipping tab

You'll need both of these for shipping quotes, and they'll automatically appear on the frontend in an additional tab beneath the **Add to cart** button:

Figure 2.5 – Additional information automatically appears when you add shipping information

> **Note**
>
> If you notice that your store is using the wrong units (centimeters instead of inches, for example), you can change this under **WooCommerce | Settings | Products General | Dimensions**. I'm going to change my units to inches and ounces.

Once you configure your shipping settings, it's important to make sure users can browse through your store and find your product. That's what taxonomies are for.

Understanding taxonomies

In addition to the **Product data** tab, you will likely want to organize your product with categories and tags. In WordPress, these are commonly known as **taxonomies** (`https://developer.wordpress.org/themes/basics/categories-tags-custom-taxonomies/`).

You can find taxonomies (**Product categories** and **Product tags**) in the sidebar:

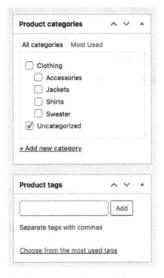

Figure 2.6 – By default, your product will be set to Uncategorized

You typically add one product category and as many product tags as you want to your product. We'll get into how to best use categories and tags to help your visitors navigate your store in the next chapter.

Once you've added taxonomies and made your product easy to discover, you need to make sure that there's an image for your product. This image will immediately tell users what they're looking at and whether they should click for more details, so it's incredibly important.

Adding images

One thing you should include on your product page is high-quality images. Images can instantly communicate what would take 1,000 words, and you can quite easily include dozens of images that engaged users will scroll through.

There are tons of articles on how valuable images can be to your store. Having multiple high-quality images that show your product in detail and context (for example, earrings shown on an ear) is crucial to your ability to sell online.

> **Note**
>
> To learn more about the importance of images, go to `https://baymard.com/blog/ensure-sufficient-image-resolution-and-zoom`.

In WooCommerce, you can add as many images as you want by using **Product image** and **Product gallery** in the sidebar:

Figure 2.7 – Product image is the default image when you see the product

> **Note**
> WooCommerce should have automatically added images to your site. For testing purposes, you can use those. I've uploaded images of my card game, Fry Thief.

The product image is the main image for the product and can be seen on the product page, in the cart, and on the shop page or any category pages.

The product gallery shows up on the product detail page. Users can see thumbnails, click or hover to zoom, and scroll through the images with their arrow keys:

Figure 2.8 – The product page after adding images

We'll learn how to rename images for **search engine optimization** (**SEO**) in the following section.

SEO tip for image filenames

We're going to cover SEO in *Chapter 4, Attracting Traffic with Search Engine Optimization,* but there is one thing you can do now to prepare your site for better optimization. As you're uploading images to your site, make sure the image filenames make sense.

Filenames such as `IMG1234.jpg` don't tell search engines anything. If you can rename your image filenames to something that provides context to search engines, this will give you a small SEO boost, as shown in these examples:

- `black-coffee-cup.jpg`
- `rainbow-umbrella.jpg`
- `48-inch-samsung-tv.jpg`

We'll continue to optimize our products later in this book, but naming your images before you upload them will save you time.

Product description and short description

In addition to all of the product data settings, there's the main content area, which is known as **Product description**:

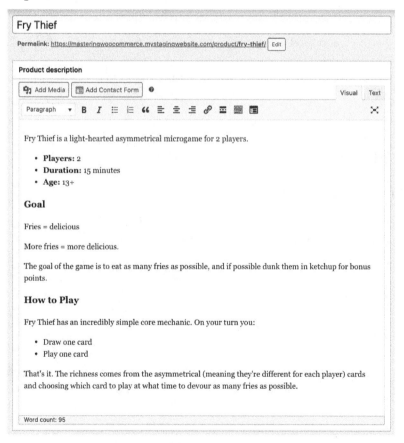

Figure 2.9 – The product description for Fry Thief

There's also a panel beneath the product data for **Product short description**:

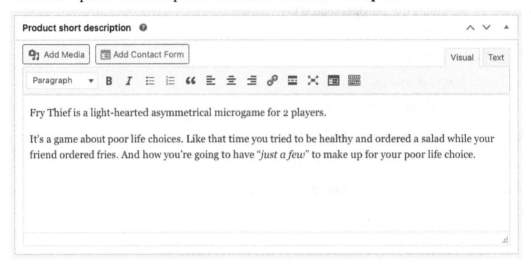

Figure 2.10 – The Product short description area shows a very brief description of your product

Both of these appear on the product page. The short description is a 1 to 2-length sentence description of the product and appears above the fold on most themes and browsers.

The product description is where you put every detail about your product. Most users won't read this but users who do will want to see complete information. Feel free to add paragraphs of content. Break it up with headings, bold phrases, images, and bullet points for readability.

A fully configured simple product

Once you've filled out all of the fields, publish your product and take a look at the resulting product page. The little pieces of information make a compelling product page:

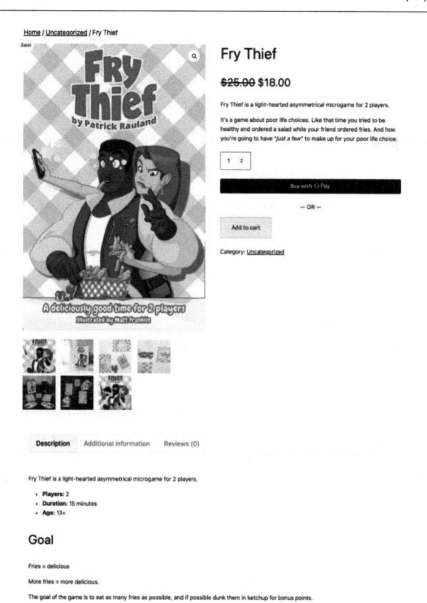

Home / Uncategorized / Fry Thief

Sale!

Fry Thief

~~$25.00~~ $18.00

Fry Thief is a light-hearted asymmetrical microgame for 2 players.

It's a game about poor life choices. Like that time you tried to be healthy and ordered a salad while your friend ordered fries. And how you're going to have *just a few* to make up for your poor life choice.

| 1 ⬍ |

Buy with G Pay

– OR –

Add to cart

Category: Uncategorized

Description   Additional information   Reviews (0)

Fry Thief is a light-hearted asymmetrical microgame for 2 players.

- **Players:** 2
- **Duration:** 15 minutes
- **Age:** 13+

Goal

Fries = delicious

More fries = more delicious.

The goal of the game is to eat as many fries as possible, and if possible dunk them in ketchup for bonus points.

How to Play

Fry Thief has an incredibly simple core mechanic. On your turn you:

- Draw one card
- Play one card

That's it. The richness comes from the asymmetrical (meaning they're different for each player) cards and choosing which card to play at what time to devour as many fries as possible.

Figure 2.11 – A fully configured simple product

Virtually, every store needs to create simple products, and you now know how to configure the essential settings.

Now that we've looked into the settings for a simple product, let's take what we've learned, change a couple of fields, and give users a few choices on the product page.

Variable products

Simple products such as coffee cups and umbrellas are great, but many products have some variation. Posters come in multiple sizes, phone cases come in multiple colors, and clothes come in sizes and colors. For these types of products, we have variable products (https://woocommerce.com/document/variable-product/).

We're going to build on what we learned in the previous section and look into a few concepts specific to variations:

- **Attributes**: These describe our products – for example, size or color

- **Variations**: These are options our users select – for example, a small T-shirt

- **Images for variation**: These are uploaded differently than simple products

- **Multi-attribute products**: These are for configuring multiple attributes in the backend (for example, size and color)

- **Troubleshooting variations**: This is for what to do when something isn't working

The first thing we have to do is select **Variable product** at the top of the **Product data** panel:

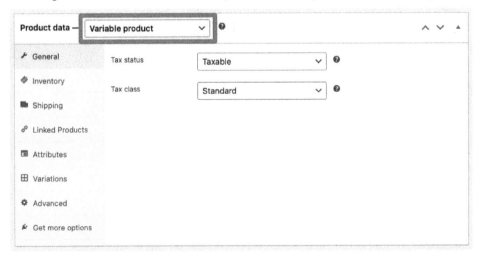

Figure 2.12 – Selecting Variable product from the dropdown at the top of the Product data panel

In many ways, variable products are configured exactly like simple products but the information in the backend is rearranged slightly.

Attributes

To be able to select an option for your product, WooCommerce has to know what the options are. These options are called **attributes** and they can be found by going to **Product data | Attributes**:

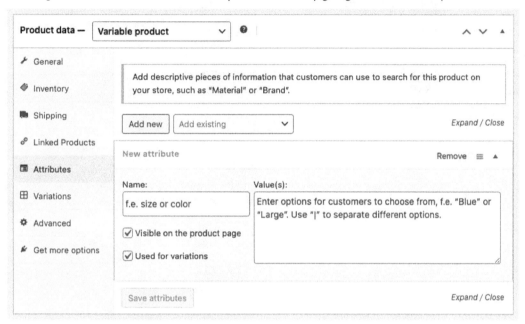

Figure 2.13 – You can create attributes directly on the edit product page

There are two types of attributes:

- **Custom product attribute**
- **Global attribute**

Functionally speaking, they're almost identical. The biggest difference is that global attributes can be reused between products, whereas custom product attributes have to be created for each product, which means more typing on your end.

Another difference is that global attributes can be used for filtering. So, if you have filters on the shop page to help users navigate your products, you can filter by specific attributes; this can only be done with global product attributes. On the other hand, if you have a product with unique attributes that no other product will have, then using custom product attributes will be more convenient.

Note

If you choose to import sample products during the welcome wizard, WooCommerce automatically imports two global attributes: **Color** and **Size**.

You can skip *Step 2* in the following instructions if you only want to edit those attributes.

1.   Let's add a global product attribute under **Products | Attributes**.

2.   Add a **Name** value for the attribute. The attribute is the name of the categorization, such as colors, sizes, and cuts – not red, small, or v-neck.

For my store, I have something more unique. Board gamers frequently like to browse games by the number of players. So, I'm adding a global attribute for the number of players:

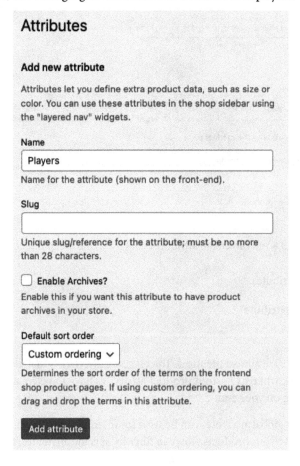

Figure 2.14 – Adding a new global attribute

1. Then, click on **Configure terms**. This will let you configure individual values.

2. Now, go ahead and add your values – for example, *Red*, *Blue*, and *Green*:

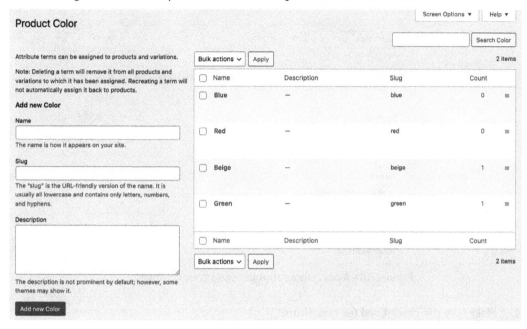

Figure 2.15 – Adding new options to a global attribute

Now, let's take a look at variations for the products.

Variations

Now that we've added attributes, we can use them to create variations. To do so, follow these steps:

1. Refresh your edit product page. Then, go to **Product data | Attributes**.

2. Select your attribute from the drop-down menu and click **Add**.

3. Now, you can select your values. If you have clothes that generally come in small, medium, large, and extra large, you might have a particular item that only comes in small and medium. In that case, just select small and medium:

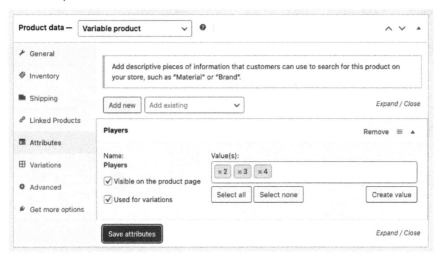

Figure 2.16 – Adding three attribute options for Fry Thief

4. Make sure you check **Used for variations**.

5. Click **Save attributes**.

6. Click on the **Variations** tab in the **Product data** panel. From here, you can add individual variations. There are two choices: **Generate variations** and **Add manually**. I recommend starting with **Generate variations** as this will create each variation for you automatically:

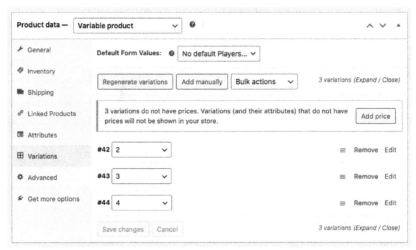

Figure 2.17 – Three variations created automatically

If you need to delete some variations, now is the time to do this. To do so, click **Remove**. I'm going to do this for my product. I have a two-player version of the game and a four-player version of the game. I'll remove the three-player version since that doesn't exist.

Editing individual variations

Once you have a variation, you can click on it and see per-variation settings. Fields such as **Regular price ($)**, **Sale price ($)**, **SKU**, and **Stock quantity** are all very easy to customize via variation:

Figure 2.18 – Editing a specific variation of a product

> **Note**
> You can add as many variations as you want, although for maintenance purposes, as well as customer clarity, I try to limit the size of the catalog.

When you're done with each variation, make sure you click **Save changes**; when you're done editing the product, click **Update**.

I've gone ahead and created two variations. Here's what they look like on the frontend:

Figure 2.19 – The two-player version of Fry Thief

Here's the same product once we've selected a different option from the drop-down menu:

Figure 2.20 – The four-player version of Fry Thief

You've probably noticed a few differences between these two versions of the product page. First are the different product images – don't worry, we'll cover that in the next section:

- Variations have different prices.

- Units in stock.

- The extra **Description** area for the four-player version. This field is useful if there are differences you need to call out for specific versions – for example, XXL includes extra buttons.

Images for variations

Before we move on, I want to highlight one aspect that's a little confusing about variations. Each variation can have a unique product image. To make this a reality, follow these steps:

1. In the backend, under **Variations**, you can select an image:

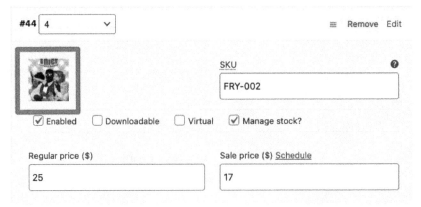

Figure 2.21 – Selecting the image icon in the top left of each variation

2. Select an image for each variation.

3. Then, on the frontend, select that variation via drop-down menus; you'll see the image replace the main product image.

This is one of the most useful application variations as you can show people what they're going to get with each variation.

> **Note**
> Don't just tell them you have a burgundy option – show them the burgundy option.

Multi-attribute variations

Some products have multiple attributes. A good example is clothing, which could come in different colors and sizes.

To add multi-attribute variations, you can follow the steps provided in the *Attributes* and *Variations* sections, except instead of adding *one* attribute, you add *multiple* attributes and make sure they're all used for variations.

With clothing, which has a different image for each color, we can create a variation for each color; for the size, we can set **Any Size**:

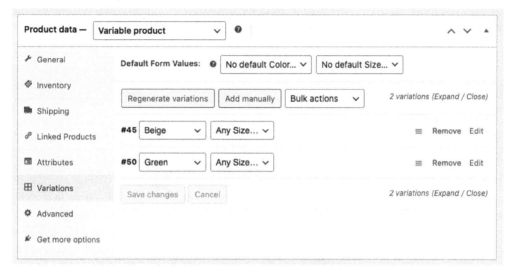

Figure 2.22 – Here, we have two variations – Beige (in any size) and Green (in any size)

> **Note**
>
> Most new store owners think they need to create a variation for each possible option. Don't do that – it takes so much time to manually set and update prices. Only create a variation when you need to set different settings – for example, a different image for each color or a different price for XL sizing. In the preceding example, if we didn't have different images, I'd use "Any Size" and "Any Color" as my only variations.

When a user selects the size and color, the main image will be replaced and users can add the item to their cart.

Troubleshooting variations

If you ever notice that you don't see the **Add to cart** button on the frontend after selecting options in drop-down menus, there are two likely culprits. The first is there's something wrong with your site, and the second is you didn't set a required field on your variation, such as the price.

If you've double-checked your variation settings and everything is fine, then it's likely some issue with your site. Since the variations functionality uses JavaScript, it's often a JavaScript error that's preventing product variations from working correctly. There's likely an error in your theme or one of your plugins.

Using a test site, disable all of your plugins and switch to a default theme; you should see the issue go away. Then, switch your theme back on and see whether the issue persists. If it doesn't appear, turn on your plugins one by one until you see the issue. By doing this, you'll know which plugin is responsible and you can reach out to the developer for support.

Now that you know how to create variable products, it's time to explore how you can sell digital products.

Digital products

Many store owners sell physical products such as shirts and mugs. But you can also sell downloadable files and virtual products, such as memberships. For this, WooCommerce lets you modify a product and give it additional settings.

Digital products aren't a separate type of product (such as simple or variable) – they're a modifier: something you can add to your simple or variable product. The new fields will appear in slightly different places, depending on the type of product. I'll be showing a simple product with digital fields. If you want to create a variable product with digital fields, the settings will be almost identical but under the **Variation** settings instead of the **General** tab.

Defining digital products

Before we change our settings, we need to define some terms. WooCommerce uses specific terms with specific meanings. You can make a product virtual and/or downloadable:

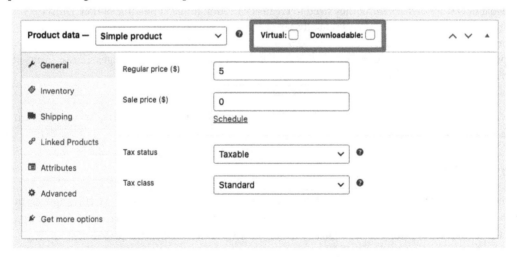

Figure 2.23 – You can select Virtual and/or Downloadable

Let's see what these options mean:

- **Virtual** means the product has no physical presence. Therefore, it won't be shipped and the **Shipping** tab will disappear. If an order exclusively contains virtual products, the customer will skip the shipping section of the checkout.

- **Downloadable** means the product has downloadable files. There will be additional fields for the downloads. The customer will automatically be emailed these files after payment is complete.

Downloadable but not virtual

Something I get asked all the time is, *Can a product be downloadable but not virtual?*

Yes! That's exactly why these are two different settings. A good example would be ordering both a physical book, which will be shipped, and an immediate download. In that case, you need to deliver files (downloadable) and you need shipping details (not virtual).

Here are some other good examples:

- Installation instructions

- FAQs

- Bonus content

Configuring virtual and downloadable products

Configuring a virtual product couldn't be easier – just check the **Virtual** setting at the top of the **Product data** panel. That's it!

Downloadable products are a little more work. Let's start by checking the **Downloadable** setting at the top of the **Product data** panel.

Under the **General** tab of the **Product data** panel, you'll see a few new fields:

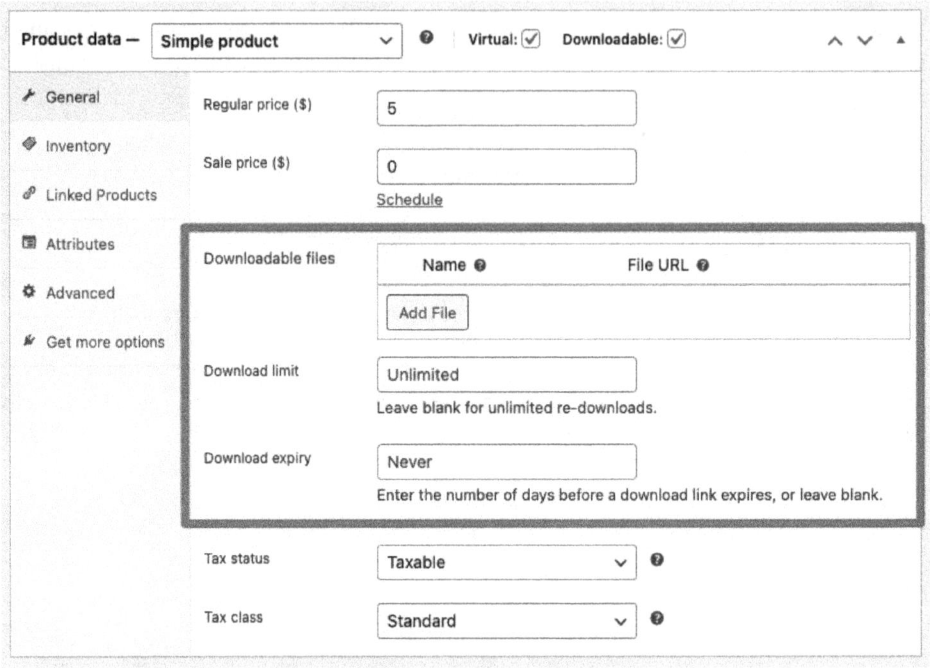

Figure 2.24 – The download settings appear after you select Downloadable

You can add as many files as you want. Give each a name and link to the file or upload a file.

Are downloadable files safe?

Some store owners worry about uploading files to their sites. WooCommerce uses a few different technologies to prevent users from pirating your files (`https://docs.woocommerce.com/document/digital-downloadable-product-handling/`).

Additionally, any files you upload will be hidden from search engines automatically.

For the most part, it isn't worth a ton of effort to prevent piracy. If a user wants to pirate your content, they'll find a way to do so. But you can always limit downloads and add an expiry time.

One way to protect PDFs is with PDF stamping (`https://woocommerce.com/products/woocommerce-pdf-watermark/`). This embeds personal user information in the PDF (in very small, hard-to-notice print). This way, you have some idea of who shared the original document. This PDF stamping product is a premium plugin, but there are also free solutions on the market.

Large downloadable files

Most web servers have enough space to host files such as desktop wallpapers, PDFs, and images. But if you have a lot of files or if you have large files, then you'll want to find a host for these files. This will speed up your site and could reduce your hosting costs.

Amazon S3 is one of the best tools for sharing static assets such as PDFs, images, and media files. And there's a fairly inexpensive plugin from WooCommerce called Amazon S3 Storage that integrates Amazon S3 with your store (`https://woocommerce.com/products/amazon-s3-storage/`).

You can also link to Dropbox or any other file storage program you want, although those methods are slightly less secure.

Accessing downloads

Once a user has purchased a downloadable product, they'll see a link to download the file(s) on the **Order received** page:

Order received

Thank you. Your order has been received.

ORDER NUMBER:	DATE:	EMAIL:	TOTAL:
55	July 28, 2023		$0.00

Downloads

Product	Downloads remaining	Expires	Download
3D Print File for Box Insert for Fry Thief	10	Never	frybox-v2

Order details

PRODUCT	TOTAL
3D Print File for Box Insert for Fry Thief × 1	$0.00
Subtotal:	$0.00
Tax:	$0.00
Total:	$0.00

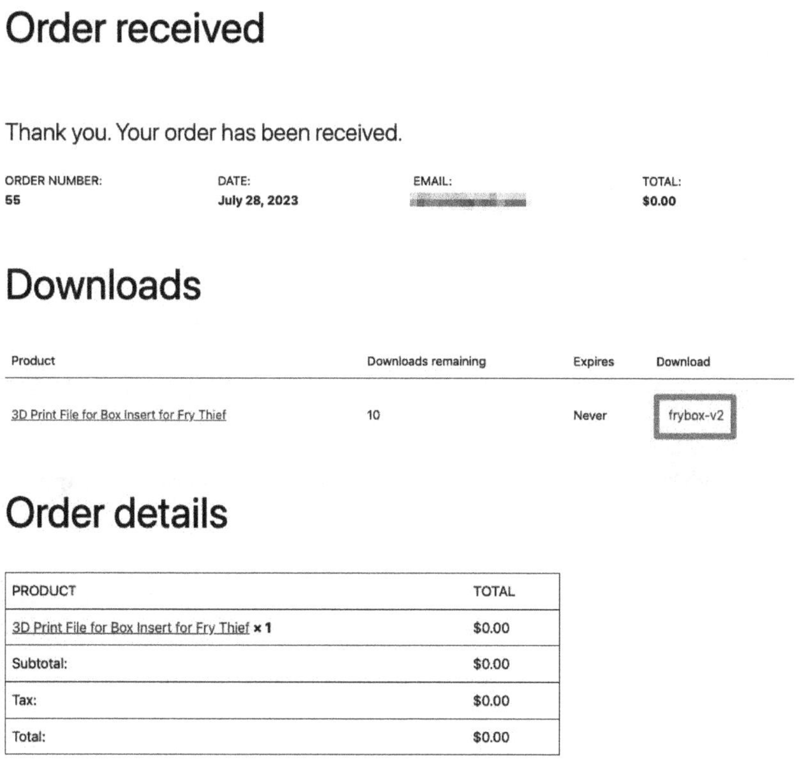

Figure 2.25 – Downloadable files can be downloaded on the Order received page

They'll also see the information in an email from WooCommerce, as well as under **My Account | Downloads** on your site. So, there are multiple places where users can find their downloads.

Now that you know how to set up a variety of products in WooCommerce, let's look at how you can sell a collection of products.

Product bundles

One of the best ways to sell more, or help your clients sell more, is to bundle your products. In WooCommerce, there's a freeway that's built-in and an extension that makes it a much easier process. We'll look at both.

Grouped products

In addition to simple products and variable products, you can also make a **Grouped product** offering:

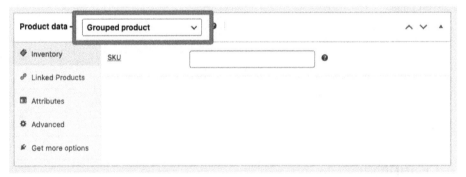

Figure 2.26 – Grouped products are built into WooCommerce

A grouped product is a collection of products that a user can select from one page. A good example of this would be products on a special sale.

When you make a product a grouped product, this will remove the **General** and **Shipping** tabs.

To add products to a grouped product, go to the **Linked Products** tab; notice the new **Grouped products** field. This field autocompletes previously created products:

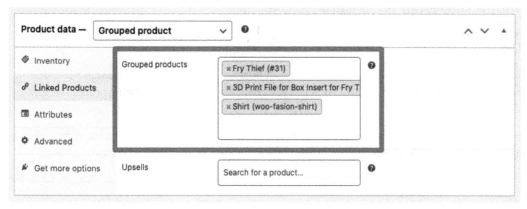

Figure 2.27 – Selecting products to be part of the group

Add a few to your grouped products and then take a look at the frontend. It can be seen in the following screenshot:

Figure 2.28 – A grouped product on the frontend

I find **Grouped product** pretty lackluster. It's probably fine for a brand-new store but most stores will want something a bit more robust – something that lets you bundle products into one package instead of a system where you are just highlighting products.

Product bundles

WooCommerce offers an extension called Product bundles (`https://woocommerce.com/products/product-bundles/`) that lets you create a group of your existing products and sell them as a group.

Once you install the extension, you'll see a new option in the drop-down menu on the edit product page. Select **Bundled Products** and you'll see a new tab appear in the **Product data** panel:

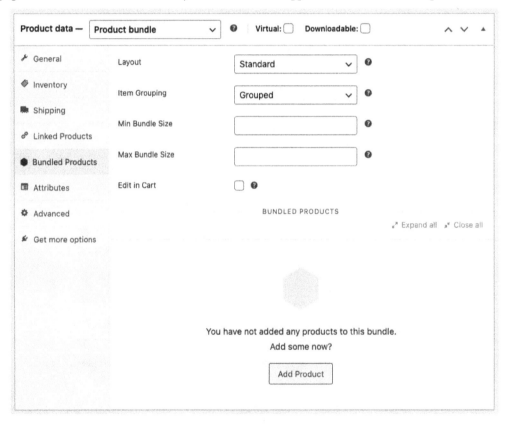

Figure 2.29 – The Bundles Products tab

On this tab, there are several new settings:

- **Layout**
- **Item Grouping**
- **Min Bundle Size**
- **Max Bundle Size**
- **Edit in Cart**

At the bottom, you can add products to the bundle. The one setting you might want to change is the **Layout** field. This will change how the bundled products appear on the product page. All of the options are useful but for our store, I'm going to select **Grid**.

We'll look at the **Min Bundle Size** and **Max Bundle Size** settings later in this chapter.

Click **Add Product** and use the auto-suggest field to add a product to your bundle. Once you've added a product to the bundle, you'll see options just for that product:

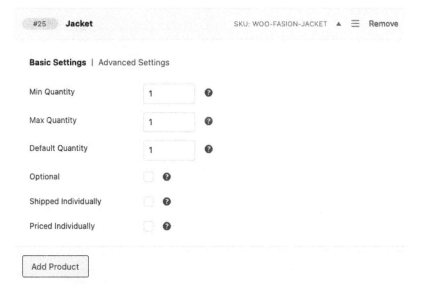

Figure 2.30 – Adding a product to a bundle

The defaults for this plugin are pretty smart and you don't have to make any changes.

In my case, I want to sell a bundle of three products together. I'll add the rest of the products, add a price under the **General** tab, and then view the frontend:

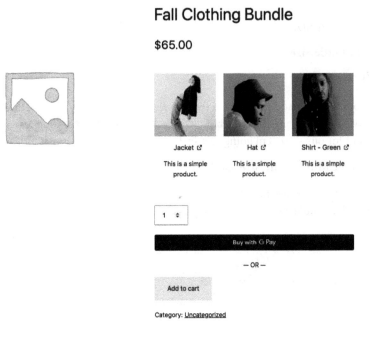

Figure 2.31 – A bundle product on the frontend

This page looks good. The only thing that I'd add to a real store is a group shot of all of these products being used.

In the next section, we will learn about configurable bundles.

Configurable bundles

I like simple bundles but you can also give users options. You can let them customize the number of items in the bundle with minimums and maximums.

Let's say we want to have a bundle where we offer a T-shirt and jacket and we let users buy a hat if they want.

We can change **Min Quantity** to 0 and **Max Quantity** to 5. Then, we can enable the **Optional** and **Priced Individually** checkboxes:

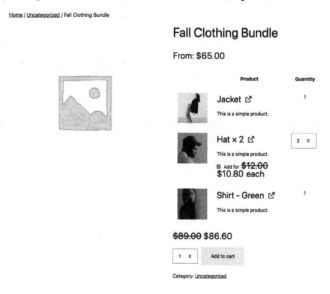

Figure 2.32 – Adding optional products to the bundle

When we check **Priced Individually**, a new **Discount** field appears. We can use this to provide a discount on this product for being in the bundle. I'll give a 10% discount.

This will work great, but we can do more for the appearance. On the **Bundles** tab, change the **Layout** field to **Tabular**. By doing this, we can see it's a bit easier to edit quantities:

Figure 2.33 – An optional product in a bundle on the frontend

If you want to use lots of optional products in your bundles, that's where the **Min Bundle Size** and **Max Bundle Size** settings come into play. You could have a jacket, a shirt, and then a choice of hats. With **Min Bundle Size** set to 3, the user would have to select one of the hats.

Product Bundles is a very powerful extension for WooCommerce. You can set up all sorts of bundles for your store.

We'll learn about product kits in the next section.

Product kits

Most stores will use bundles at some point, which is why I showed off the **Product Bundles** extension. Fewer stores use product kitting but it is still very powerful. Product kitting is where you select an item from bucket 1, an item from bucket 2, and an item from bucket 3 and put them all together.

This can be achieved with WooCommerce with an extension called **Composite Products** (`https://woocommerce.com/products/composite-products/`). But since a relatively small number of stores have products like that, I don't want to use valuable page space; instead, I recommend that you refer to the official product page and documentation.

Now, let's move on to the only product type that brings in recurring revenue.

Subscriptions

The cost to get a new customer (also known as the acquisition cost) is typically high. Many eCommerce stores only make a minuscule profit on the first transaction because you often spend money attracting that customer with ads, social media, content, tradeshows, and so on.

With acquisition costs being high, a lot of store owners love recurring payments, where customers pay every week, month, or year. That means they may only make a little profit on the first order but the subsequent orders are much more profitable.

WooCommerce *Subscriptions* (`https://woocommerce.com/products/woocommerce-subscriptions/`) is one of the most powerful subscription products on the market with all sorts of advanced features, such as the following:

- Pausing subscriptions
- Prorating subscriptions
- Synchronized payments

Creating a subscription product

To create a subscription product, you have to purchase and install the WooCommerce Subscriptions extension. Once you do, you'll see a new option on the edit product page.

Select **Simple subscription**, although note that you can also choose **Variable subscription**, which is similar to variable products. When you select a subscription product, you'll see new fields:

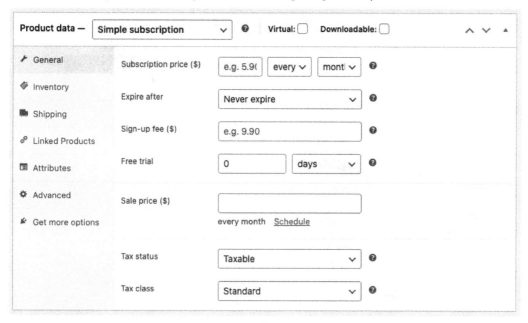

Figure 2.34 – Creating a simple subscription product

As you can see, there are a ton of options. You can select the following:

- How often you want to bill people (weekly, bi-weekly, monthly, yearly, twice annually, and so on)
- When the subscription should expire – if ever
- Whether there's a free trial or whether there's an additional sign-up fee
- Of course, a price

When you take a look at the frontend, you'll see that the interface has changed to show all of our options to the visitor:

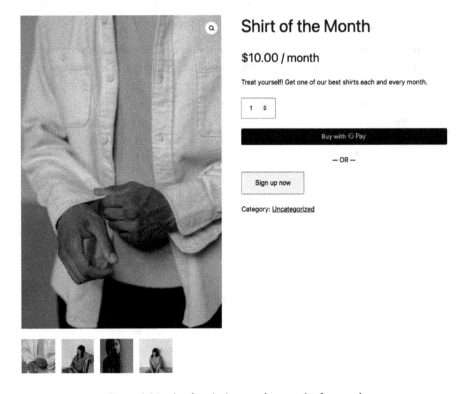

Figure 2.35 – A subscription product on the frontend

If you offer a product that's purchased primarily through a subscription, these settings should suffice.

We'll add a recurring payment option for the users in the following section.

Adding a recurring payment option to a product

If you've been to Amazon.com or other big online retailers, they will often offer a "subscribe and save" option for one-off purchases.

This lets users buy a product one time or if it is something they buy regularly, they can subscribe and save some money – and the retailer is happy with consistent sales.

You can do this with your WooCommerce store with *All Products for WooCommerce Subscriptions* (https://woocommerce.com/products/all-products-for-woocommerce-subscriptions/), which is another paid extension by WooCommerce.

You can turn any product into a subscription, as shown in the following screenshot from the WooCommerce website:

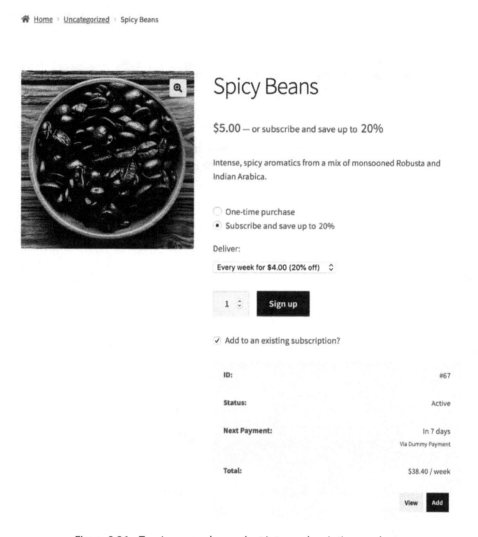

Figure 2.36 – Turning a regular product into a subscription product

We'll get started with subscription settings in the following section.

Subscription settings

Subscriptions are complex and have their own settings page. Under **WooCommerce | Settings | Subscriptions**, you can find a ton of settings.

Manual versus automatic renewals

If you live in the US, Canada, or Europe, you should have access to modern gateways that provide your store a token so that you can charge credit cards without having to save the credit card numbers directly.

> **Note**
>
> WooCommerce has a list of subscription-compatible gateways on their site. Here is the link if you want to take a quick look: `https://docs.woocommerce.com/document/subscriptions/payment-gateways/`.

A modern payment gateway is all WooCommerce Subscriptions needs to automatically charge someone's card. But if you're having a hard time finding the right payment gateway, you might not be able to get credit card tokens and bill someone regularly automatically. In that case, there's a feature called **Accept Manual Renewals** that you can enable:

Retry Failed Payments	☐ Enable automatic retry of failed recurring payments
	Attempt to recover recurring revenue that would otherwise be lost due to payment methods being declined only temporarily. <u>Learn more</u>.
Renewals	
Manual Renewal Payments	☐ Accept Manual Renewals
	With manual renewals, a customer's subscription is put on-hold until they login and pay to renew it. <u>Learn more</u>.
Auto Renewal Toggle	☐ Display the auto renewal toggle
	Allow customers to turn on and off automatic renewals from their View Subscription page.
Early Renewal	☑ Accept Early Renewal Payments
	With early renewals enabled, customers can renew their subscriptions before the next payment date.
	☐ Accept Early Renewal Payments via a Modal
	Allow customers to bypass the checkout and renew their subscription early from their **My Account > View Subscription** page. <u>Learn more</u>.

Figure 2.37 – Enable Accept Manual Renewals if your payment gateway can't save credit card tokens

Users who have manual renewals will receive an email and will have to click a link and pay for an order every month. It's much more work for the end user and is only designed for stores that can't get access to modern payment gateways.

If you have a modern gateway, I suggest leaving this feature off.

Subscription switching

If you have a bunch of related subscriptions, you might want to enable subscription switching. For example, let's say that every month, you deliver a 24-pack of soda. If the subscriber starts to overflow with soda rather than cancel, give them a smaller soda pack and a smaller discount so that they stay subscribed and can upgrade in future months:

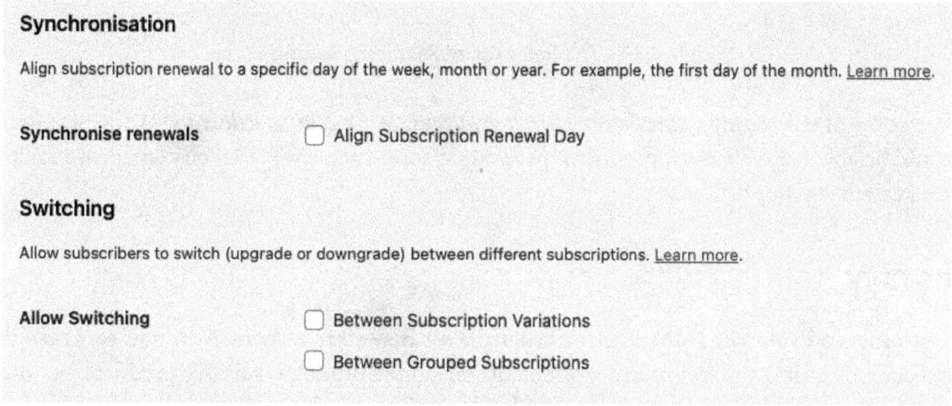

Synchronisation

Align subscription renewal to a specific day of the week, month or year. For example, the first day of the month. Learn more.

Synchronise renewals   ☐ Align Subscription Renewal Day

Switching

Allow subscribers to switch (upgrade or downgrade) between different subscriptions. Learn more.

Allow Switching   ☐ Between Subscription Variations
                                 ☐ Between Grouped Subscriptions

Figure 2.38 – Synchronizing subscriptions is handy if you send products every week or month

I'm a big fan of enabling subscription switching between variations, but you can also enable them between a group of subscription products.

Synchronization

Some subscription stores do everything on a schedule. As an example, I subscribed to *The Simple Jar*. They deliver food weekly each Monday.

They get all orders on Thursday, prep them on Sunday, and deliver Monday morning. For businesses like this – for example, monthly loot boxes – that have a particular schedule, synchronizing subscriptions is hugely helpful. Customers can subscribe at any point and they'll get the next shipment, at which point their payment will be postponed until the next billing cycle.

Check the **Align Subscription Renewal Day** checkbox and then follow the instructions under **Learn More** to enable this feature.

Retrying failed payments

Lastly, there's one other feature that's worth mentioning. Credit card numbers change and payments fail. Subscriptions have a feature that (under some conditions) will retry failed payments (`https://docs.woocommerce.com/document/subscriptions/failed-payment-retry/`). This gives credit cardholders time to pay off their balance or get their new credit card in the mail.

Once you start getting the occasional failed payment, ask your audience if this is something they'd want. If so, you can enable it in your settings:

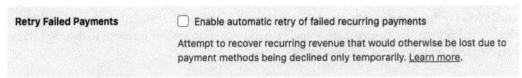

Figure 2.39 – The Retry Failed Payments setting

These are some of the settings every store owner and manager needs to know for WooCommerce. You should be able to create most product types and do so in such a way that customers are excited to see and purchase your product.

Summary

Listing products on your site is the first major hurdle for new store owners. Now that we know the pros and cons, as well as the technical requirements, of simple products, variable products, bundles, and subscriptions, as well as how to add downloadable files to each of them, we can start adding products to our site.

Once we added products to our store, we turned our site from a brochure site into an online catalog. That's halfway to a full online store.

Now that you know how and why to add different types of products to your store, in the next chapter, we can look into organizing and presenting these products to users so that they can find the right product and add it to their cart.

3
Organizing Products

Once you've added products to your store, it's time to start thinking about the organization of your store. WooCommerce automatically lists products on your Shop page and if you're new to e-commerce, you might decide that's enough and stop there.

But there's actually a lot you can do to improve the organization of your store, and this organization can have a huge impact on the revenue your store generates.

As an example, in 2017, WooCommerce updated the category structure on its site. It reorganized 17 top-level categories into 7 top-level categories with subcategories.

This one change improved the conversion rate of anyone landing on a category page by 20%. That's a *massive* improvement, and what's great is that this helps both the store owner by getting more revenue and the user by making it easier to find the information they need.

Having a great structure for your store will also naturally help with your search engine optimization and drive more people to your category pages. Hopefully, now you understand how powerful it can be to properly organize your products.

In this chapter, we're going to cover the various ways you can organize your products, including the following:

- Categorizing and tagging
- Optimizing product archive pages
- Adding product filters to your Shop page
- Using product blocks through your site

Let's start with categorization and tagging.

Technical requirements

In this chapter, we're going to install the Storefront theme (`https://wordpress.org/themes/storefront/`). This is the official theme from the WooCommerce team and is an excellent starting point for any store.

Later in this chapter, we'll install the **Yoast SEO** plugin (`https://wordpress.org/plugins/wordpress-seo/`). This plugin helps you improve your **search engine optimization** (**SEO**) across your site. We'll use it to help us improve the SEO on our category pages.

Categorizing and tagging

First, let's talk about what's possible with two of the commonly used WooCommerce organizational tools: **categories** and **tags**.

Each product in WooCommerce can have categories and tags. Categories tend to be hierarchical, while tags have a flat structure. It's a best practice to have one category and several tags.

Here's an example of a product's category, and you can add tags right below it:

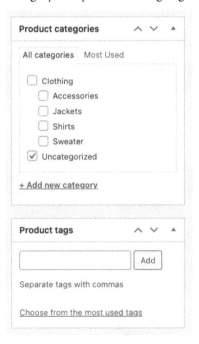

Figure 3.1 – The built-in product categories and product tags

You might notice that you *can* add multiple categories to a product, but that doesn't mean you *should*. I recommend you only use *one* category for each product.

WooCommerce with the Storefront theme and a few other themes has a really nice breadcrumb feature that shows users where they are in the catalog. Breadcrumbs, which we'll talk about later in this chapter, don't work well with multiple categories.

You can see the breadcrumbs in the following screenshot:

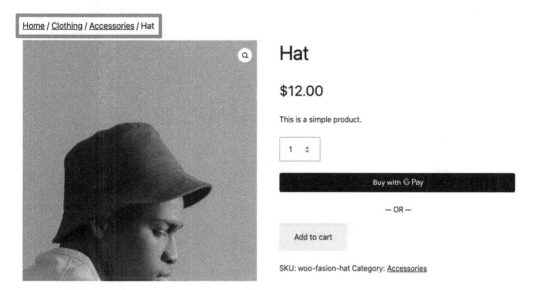

Figure 3.2 – Breadcrumbs help users backtrack through categories

The breadcrumb feature on the product page lets users easily navigate to parent categories or all the way back to the Shop page.

To make categories useful to your users, you'll want to make sure they're clearly named and understood and there's no ambiguity. To do that, I use a strategy called **mutually exclusive and collectively exhaustive**. Let's find out more about it.

Mutually exclusive and collectively exhaustive

To help users navigate your store, you have to have a well-designed category structure. The phrase I always use is that categories have to be mutually exclusive and collectively exhaustive.

Mutually exclusive means categories shouldn't overlap with each other. So, each item should fit clearly into exactly *one* category.

Collectively exhaustive means there should be a category for every type of product you sell. Avoid categories such as *Other* or *Miscellaneous* since shoppers rarely click on them, and they're essentially a digital junk drawer.

So, for a clothing store, a good set of categories would be as follows:

- T-shirts

- Hoodies

- Accessories

A bad set of categories would be as follows:

- Smiley faces

- Blue apparel

- V-necks

- Miscellaneous

All of these categories overlap with each other, and store owners will be tempted to put the product in multiple categories, which is confusing to users. As a store builder, you'll want to make sure that none of your categories overlap with one another and that you'll make a structure that makes sense to users, which will help your store to make more sales.

Once you have a solid set of categories, you'll want to add a few tags to your product.

Tagging products

Now that we have a system for categorizing products, we need a system for tagging products. You may be tempted to copy the category structure.

A store in my local area that happens to use WooCommerce has a category and tag structure that looks like the following screenshot:

Dark Chocolate REGULAR Caramels

$16.95

The ▒▒▒▒▒ ▒▒▒▒ ▒▒▒▒▒ original-recipe caramel hand-dipped in pure dark chocolate, also available in milk chocolate. 16-piece box. All of our caramel is made without corn syrup, full ingredients here. One minute podcast review here

| 1 | Add to cart |

Categories: Assorted, Best Sellers, Caramels, Gifts for Him Tags: all-natural dark chocolate caramels, best caramels, caramels, corn syrup free, dark caramels, dark chocolate, dark chocolate caramels, gluten free caramels, handcrafted caramels, no corn syrup

Figure 3.3 – Look at all of these tags! So many to choose…which do I click?

Do you see anything confusing for the user about this structure? If I want to browse by caramel chocolates, what do I click?

- **Categories | Caramels**
- **Tags | Caramels**
- **Tags | best caramels**
- **Tags | handcrafted caramels**

This is actually a common issue. Many store owners have a tag structure that's too similar to product categories. In such cases, both users, as well as search engines, tend to get confused. You want to make sure it's 100% obvious what the user should click on. This will increase conversion rates.

What I recommend is to use one system for categories and a completely different system for tags. So, going back to clothing, I might use the following as categories:

- T-shirts
- Hoodies
- Accessories

And these would be good tags to navigate between the categories:

- Smiley faces
- Blue apparel

When your client comes to you with 100 categories and no subcategories, now you know better. Customers will find that hard to navigate, and your client's store might fail. You now know how to clearly categorize and tag your store to help users find the right products so the store can make more money.

Once we have that structure in place, it's time to optimize the product archive pages.

Optimizing product archive pages

Now that we have intuitive categories our users can browse, let's optimize these category pages. We can make them much easier to read both for the user and also for search engines. We're going to do the following:

- Write descriptions for the product category pages
- Clean and simplify our URLs so they're easy to understand at a glance

When we're done with this section, users will understand the product archive pages (category pages) and since search engines better understand your store, you should also get more inbound traffic.

Writing descriptions for product categories

One of the most important things you should do is add a description to your product category pages. To do that, follow these steps:

1. You can do this in WooCommerce in the backend through **Products | Categories | {{pick a category}}**:

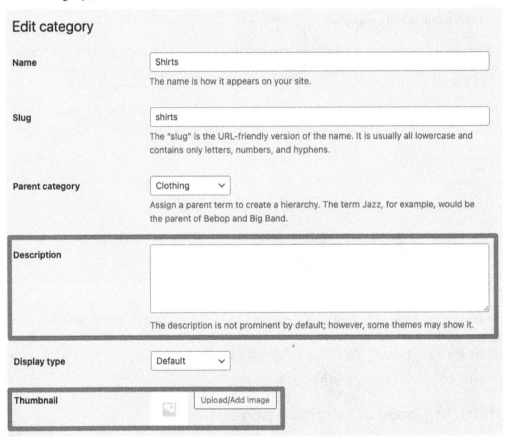

Figure 3.4– You can customize several details on this page, including the category description and the thumbnail

2. From here, you can write a description and upload a thumbnail for your category, as shown in this screenshot:

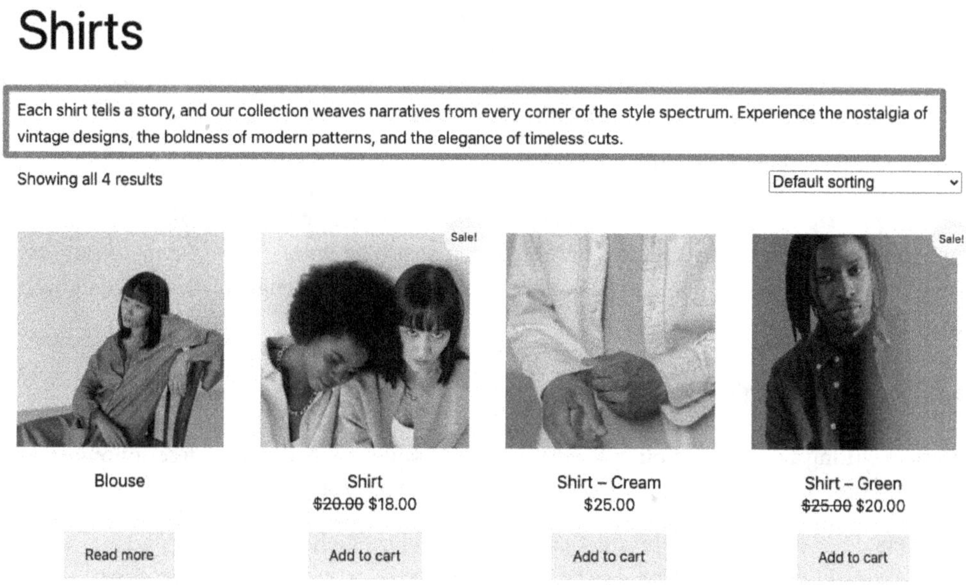

Figure 3.5 – Adding a short description to a category page

In most WordPress themes, you can see the category description above the products. However, different themes can display this information in different places and some themes hide this information altogether.

> **Note**
>
> Unfortunately, many themes don't display a product category thumbnail. Even though not all themes will display the thumbnail, this image can still be selected to be shared on social media, so you might want to select an image in case this page is shared on social media.

If you haven't yet decided on a theme for your store, I recommend Storefront (`https://wordpress.org/themes/storefront/`). You can find out more about themes in *Chapter 10*.

Let's see how to add proper context for the products in the following section.

Context matters

One thing your product category description can do is provide context. Let's say you're selling mice for laptops and computers. If you have a product category called Mice and you have products called Red Mouse and Blue Mouse, search engines won't know whether you're talking about:

- Mice – the animal
- Mice – the computer accessory

The category description can provide that context. If you're selling computer mice, make sure to mention "computer" as well as semantically related words such as "mouse pad," "smooth scrolling", and so on in your description. Semantically related words help search engines disambiguate words with multiple meanings.

Meta-description for categories

In addition to writing a good description for the page itself, you can write a meta-description that will catch someone's eye so they click on your page in the search engine results page.

To add a meta description to your product category, you'll have to install an SEO plugin. I usually install the **Yoast SEO** plugin (https://wordpress.org/plugins/wordpress-seo/), which is available for free at https://wordpress.org:

Figure 3.6 – Yoast SEO makes it easy to configure your Meta description

With this plugin, you can write a meta description that will be displayed on the **Search Engine Results Page (SERP)** and you'll see a preview within your admin.

Search engines bold any searched terms on the SERP. So, if the user searches for `vintages design t-shirt`, then `vintage design` in my description will be bold and will draw their eye.

You can also customize the SEO title. `Shirts Archives` isn't that helpful. We can probably remove `Archives` so that it's clear this is a list of t-shirts.

A pretty category description

As shown in the following screenshot, once you install Yoast SEO, you might have noticed that your product category description field is now a rich text field where you can add bold, italics, images, headings, and so on.

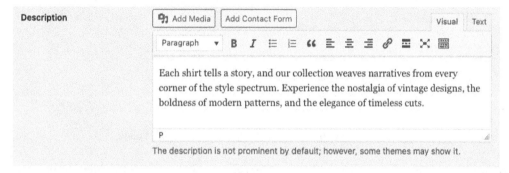

Figure 3.7 – The category description is now a rich text editor

The default category description used to be plaintext. It was just an empty text box where you could only type characters. Now we have lots of formatting options, such as bolding, italicizing, and adding bullets.

URLs

Something that's always important for SEO is to have clear URLs. You want to have one clear term on your product category page. So, you might want to have the following:

- t-shirt
- tshirt
- tee

But you definitely don't want this:

- t-shirt-tshirt-tee

Search engines will penalize you if they think you're spamming keywords. So, make sure you pick the most popular option for your industry and use that. If you have already installed an SEO plugin such as Yoast SEO, you can edit the **Slug** field:

Figure 3.8 – The slug is what shows up in the URL for this page

If you aren't using Yoast SEO, you can do this in WooCommerce itself under **Products** | **Categories** | **{{pick a category}}** | **Edit** and then edit the **Slug** field.

Redirects in WordPress

One thing you should know about is changing URLs. WordPress does a pretty good job automatically creating redirects. If you create a product category called T-shirts and you change it to t-shirts, the old URL should redirect to the new URL.

This is fantastic behavior unless an admin changes the URL back to the original URL. If you go from tshirts to t-shirts and then back to tshirts again, you might have an issue with infinite redirects.

Our product archive page is a lot easier to read and a lot easier for search engines to find. Now, we can look into improving the Shop page with product filters.

Adding product filters to your Shop page

Since we're organizing the products in your store, if you haven't installed Storefront, now is a good time to do so. Most themes will work with WooCommerce, but Storefront is designed to work with WooCommerce, and it's made by WooCommerce so it's a great place to get started. Whatever theme you choose, you'll want to make sure you can configure a sidebar so we can add widgets to help our users filter products.

You can do this by going to **Appearance | Themes | Storefront | Activate**.

Now we want to go to the WordPress customizer. You can get to the customizer through **Appearance | Customize** in your admin. You should see something like this:

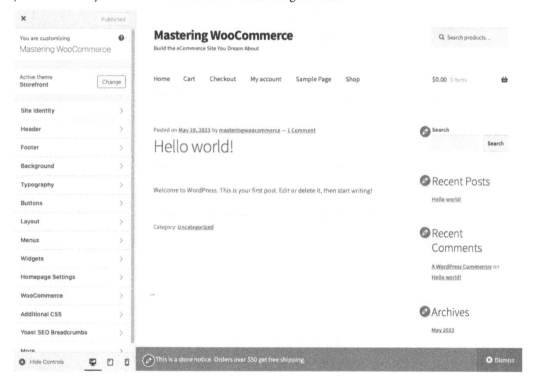

Figure 3.9– The WordPress customizer

Let's customize our Shop page. To do that, follow the steps given here:

1. Click on **Shop** from within the customizer, which will load the following page:

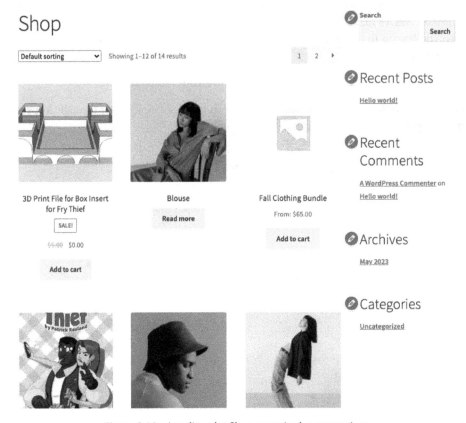

Figure 3.10 – Loading the Shop page in the customizer

By default, down the right-hand side, we see the following:

- **Search**
- **Recent Posts**
- **Recent Comments**
- **Archives**
- **Categories**

And for some blogs, these are fine. But for an online store, we want to be able to search and filter products.

1. In the sidebar of the customizer, click **Widgets | Sidebar** and you'll see the list of widgets:

Figure 3.11 – Configuring the sidebar in the Customizer

If you're an experienced WordPress user, these might look a little different than you're used to. The widgets are now controlled with Gutenberg blocks. They're still easy to use but the **user experience (UX)** is different.

2. Go ahead and hover over each widget and click the three-dot icon. Then click **Delete** to remove the widget. Do this with all of the widgets. We're going to start from scratch.

Here's how you can delete the widget:

Figure 3.12 – Deleting widgets in the Customizer

3. Now, it's time to add useful widgets to our store. Click the + icon to add a widget. In the search field, type in `woocommerce` followed by **Browse All** to see a list of widgets designed for WooCommerce:

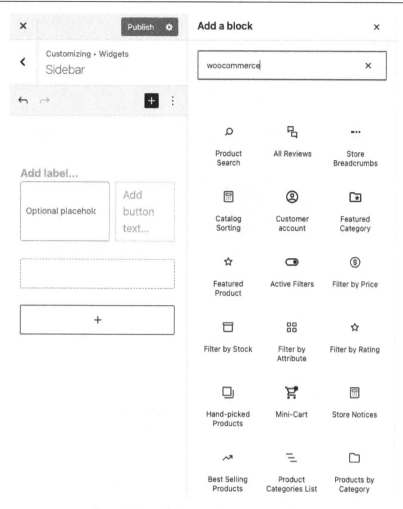

Figure 3.13 – Adding WooCommerce widgets

Now, add the following widgets:

- **Product Search**
- **Active Filters**
- **Filter by Attribute**
- **Filter by Price**
- **Filter by Rating**
- **Product Categories List**
- **All Reviews**

The widgets can be seen in *Figure 3.15*:

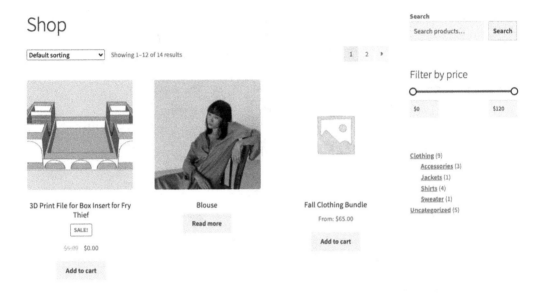

Figure 3.14 – Adding filter widgets to the Shop page

You might notice that not every widget shows up. For example, where are the product attributes? There are circumstances when WooCommerce hides widgets. Let's see how this works.

Hidden widgets

Widgets in WordPress are pretty smart, and they only show up if they're relevant. If none of your products have attributes, then you won't see the attribute filtering widget. The same thing happens with product reviews: if there are no recent reviews, then there's nothing to display.

Let's add some product attributes, such as color, to our store. You can read *Chapter 2, Configuring Products*, to see how to add global product attributes. Once you do, you can see them in the sidebar and then you can click them and start filtering the products.

Once you have a global product attribute, you can customize your **Filter by Attribute** widget:

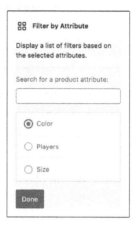

Figure 3.15 – Configuring the Filter by Attribute widget for our product attributes

You need to select one attribute. In our case, that's going to be color. But if you have multiple attributes, such as color, size, and cut, you can add multiple **Filter by Attribute** widgets.

Active product filters

I've gone ahead and clicked on a color (**Green**) and set a maximum price range. Now, with the active product filter, we see the exact parameters we set on our products:

Figure 3.16– The Active filters widget shows us what filters are currently in use

Product filters are somewhat useful for a dozen products, but as you get into hundreds or thousands of products, product filters become very helpful to your shoppers, letting them browse according to their needs.

The Active Product Filters widget lets users see those parameters and remove them. You don't technically need this widget but without it, some users will be confused about how to remove parameters.

Filters make your Shop page much easier to navigate and I strongly recommend them. Once you've made the Shop page easy to navigate, we can look into adding product blocks around our site.

Understanding product blocks

With WordPress 5.0 (Gutenberg) came blocks, and the WooCommerce team immediately started creating custom blocks for products (`https://woocommerce.com/document/woocommerce-blocks/`).

To see how powerful these blocks are let's create a new homepage. Go to the admin, then go to **Pages | Add New**.

Once the new page loads, it's time to configure the homepage.

Add a title for the homepage. I like to keep it simple and obvious by calling it **homepage**. Now let's add a few blocks to this page.

The simplest solution is to add the **All Products** block. But for my store, I want to highlight the product I'm best known for. So I'm going to use the **Single Product** block. Then select a specific product from the dropdown. I'll choose *Fry Thief*.

Figure 3.17 – The Single Product block

I also want to show off some of our swag beneath the main product. Let's add the **Products by Category** block and then select the **Clothing** category from the options available.

I want to end the page by showing some social proof, meaning testimonials and reviews from customers. Let's add the **All Reviews** block:

> **Note**
> We don't have any reviews yet, but as soon as we do they'll appear right where this block is.

1. For both of these blocks, you can customize their settings by making sure you have the sidebar open and selecting the block. I change the number of rows for the category to just 1 so it only shows 3 items rather than 9.

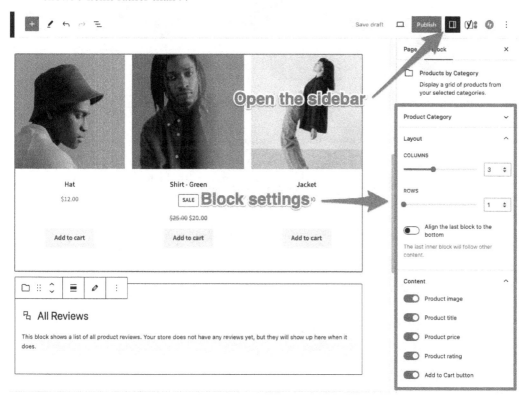

Figure 3.18 – Many blocks have powerful settings

Finally, let's add some regular blocks to add some text to the page. It's all products at the moment and we want to guide our users from section to section. Let's add the **Cover** block. You can then select an image from your media library and add a heading.

Figure 3.19 - The Cover block lets you add text over an image

2. I'll add the **Heading** block just above the clothing category, and I'll add the text Buy Some Swag

3. I'll also add the **Button** block to the bottom of the page with the text Shop Our Whole Catalog

4. I don't want our homepage to have a sidebar. I want it to be full-width. Most themes allow you to do this with templates. With Storefront, I have three templates to choose from. I personally like the **Full Width** template, which you can select in the sidebar under **Page** settings.

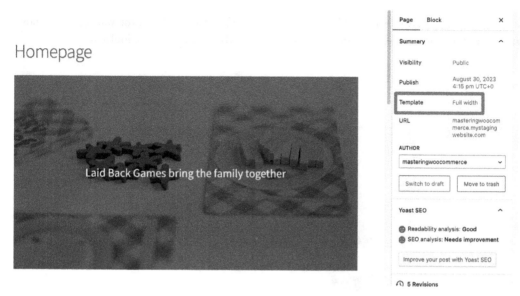

Figure 3.20 – Many themes have templates for added customization

If you want to make this page the homepage, you can do that from the WordPress Admin then **Settings | Reading | Your Homepage Displays | A static page |** {the name of your page}:

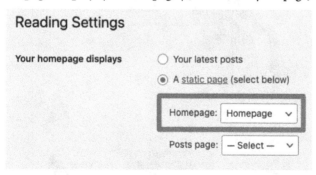

Figure 3.21 – Setting the homepage for my site

Configuring a homepage only takes a few minutes once you know what blocks you want and you know how to configure them. And, after all of that work, here's what our homepage looks like:

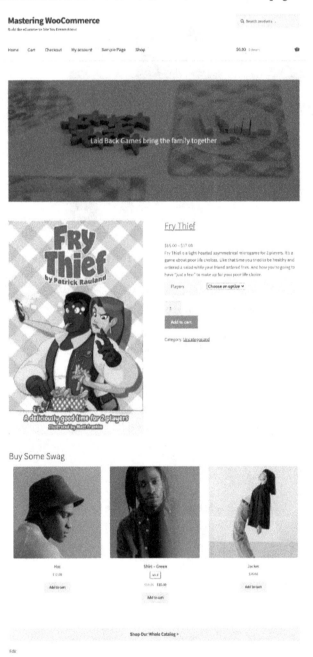

Figure 3.22 – The completed homepage

> **Note**
>
> If you want to see other blocks that WooCommerce is making, you should download and install the free extension WooCommerce Blocks (`https://woocommerce.com/products/woocommerce-gutenberg-products-block/`). As WooCommerce is developing new blocks, they'll typically release them in this extension first and if they have a large adoption, they'll move them to the core plugin.

Single product pages

With this much control, it's really easy to create great homepages. But you can also create fantastic single-product pages or product release posts.

You can combine regular blocks such as **Cover image**, **Image**, and **Paragraphs** and combine them with a WooCommerce block such as **Hand-picked Products**.

These product pages don't have to be regular web pages – they can be announcement posts or anywhere you see the modern Gutenberg block editor in your admin.

For this book, I'm going to use the default pages in WooCommerce, but it's good to know I could create my own single product pages if I wanted to make something truly unique.

Customizing product blocks

When you add a product block to your post, there are several settings you'll want to tweak:

Figure 3.23 – Configure block columns for the best layout

The first is likely to be the **Columns**. If you're highlighting a set of new products, make sure you have the right number of columns for your products. I'm only showing one new product so one column will work fine.

> **Note**
> I'm choosing to show all product information (title, price, ratings, and the **Add to cart** button), but WooCommerce is smart and leaves the rating empty until there are ratings. Once a user rates this product, this will dynamically appear.

If you need to change the products on display, you can click on **Products** under **Block Settings** and add, edit, or remove products.

Featuring a product

The best way to feature a product is with the **Single Product** block, but I expect this to change (maybe even by the time this book comes out). This block shows the product title, image, price, and an **Add to card** button, so it's pretty comprehensive.

Keep an eye on the product blocks as they're a fantastic way to make use of all of the content power built into WordPress.

Keep an eye on blocks: in general, if you go to `WordPress.org`, you'll see blocks for many features, such as testimonials, graphs, and graphics.

The value of blocks

What's so great about these new blocks is they show you what your user will see on the frontend and they display dynamic content. That means if you ever update your featured image, price, or even the title of the product, all of that will be changed wherever you use these blocks.

With some of my clients, they don't want to mention product details such as the price in a release post because the price might change and they don't want users to be confused. With blocks, you don't have to worry about that at all. Its dynamic content is visually displayed in the post editor, which is perfect for store owners.

Summary

We looked at all of the ways to organize your products so users can easily find them and search engines can send you more traffic. From very basic techniques such as adding categories and tags to more advanced techniques such as adding product filters and using product blocks in the Gutenberg editor, you have a variety of tools to help users find the right product for them.

Organizing products is a bit of art and science. It's worth doing some research before you launch your store to organize your products in a way that will make sense for shoppers. Once you start getting real-life users, you can analyze their browsing and search habits to see whether you can make improvements.

Now, we can learn how to optimize the rest of our site for search engines (a process called SEO) and learn how to attract traffic in the next chapter.

4

Attracting Traffic with Search Engine Optimization

You can have the perfect eCommerce site with thousands of products, perfect organization, convenient ways to search, easy-to-use payment options, and cheap shipping. Unfortunately, none of that will do you any good unless you can get people to your website.

Bringing people to a regular website is called **traffic generation**. In the eCommerce world, you're trying to bring people to your site to make purchases. This is called customer acquisition, and **customer acquisition cost (CAC)** is one of the main indicators of a successful eCommerce business. If you can reliably bring traffic to your site, a small percentage of that traffic will purchase products, and you'll make money.

There are techniques to build a strong eCommerce business, such as securing recurring purchases, increasing conversion rates, and lowering your product costs, but none of them matter without the initial traffic! One of the main techniques to bring in traffic is **search engine optimization (SEO)** – basically, getting search engines to recommend your site more often. So, if there's one thing you should focus on after you build your store, it's customer acquisition.

In this chapter, we're going to cover the following main topics:

- Why you should invest in SEO
- Keyword research for eCommerce
- Configure breadcrumbs to help search engines and users
- Creating and sharing an XML sitemap
- Why you should keep an eye on Google Search Console

Technical Requirements

The code files for this chapter can be found in the following GitHub repository: `https://github.com/PacktPublishing/Mastering-WooCommerce-/tree/main/Chapter04`

Why you should invest in SEO

There are many ways to bring people to your site. The following are some of the best methods:

- Paid advertising
- Social media
- Influencer marketing
- Affiliate programs
- Trade shows
- Business development
- Speaking engagements
- Public relations
- Content marketing
- SEO

While we can't delve into every strategy within the scope of this discussion, it's worth highlighting the distinctive nature of SEO and its connection to content marketing compared to other approaches.

First, we'll look at some of the most talked-about strategies, which are usually advert-related and cost money to get up and running. Then, we'll look at the slow but steady increase in visitors from SEO.

One-off marketing strategies

Most customer acquisition strategies are transactional. By that, I mean you do something once and you get a certain number of visitors. Here are some examples:

- You spend $10 on a Facebook advert and get 300 visitors.
- You go to a trade show and meet 100 people. 10 of them eventually become customers.
- You announce a new product that gets picked up by the media and you get 1,000 visitors to your store.
- You create a funny TikTok and you get 200 visitors.

Most of these require ongoing effort or money for them to bring in visitors. However, SEO and its cousin, content marketing, are different. You take the time to optimize your site or write an article once and they continually bring in visitors.

Always-on marketing

If you write a high-quality article so that search engines think you're the best result for a particular query, it can bring in dozens to thousands of visitors a month.

I've been writing about eCommerce for years on my blog (`https://www.speakinginbytes.com/`). The posts that I wrote 10 years ago are still bringing in traffic every month.

However, this process takes some time. For search engines to recognize that your site provides value, both your site and the post have to be around for a while. Search engines also rely on other sites linking to you, and of course, it takes a while to write the articles.

Let's say you can write one article a week and those articles generate only 20 visitors a month for the first several months. That's a lot of effort for a total of 80 visitors. And with a typical 1% conversion rate, that's 0.8 orders.

> Tip
> SEO and content marketing can be great but they are very slow to ramp up, so you'll want to combine them with other strategies that bring in a large number of visitors for a one-off cost.

Let's investigate which keywords are the most popular in helping with both SEO and advert-buying strategies.

Keyword research for e-commerce

Before you optimize your site, it's worth researching to know what you should optimize for. Doing keyword research helps you figure out which words or phrases people are searching for and which ones you can rank for.

Most stores open without doing any research, and they think they'll rank for hugely popular terms such as the following:

- Shoes
- T-shirts
- Chocolate
- Razor subscription

These terms are *highly* competitive. With a little research, using tools later in this section, you can optimize your site for search terms that are much easier to rank for. We'll look into the following:

- Creating a list of potential keywords
- Comparing the search volume of different keywords
- Optimizing for the best solution

Once you know how to do all of these, you'll know which terms you can try to rank for in search engines to get some free traffic.

Creating a list of keywords

The first step to optimizing your site for search engines is to create a list of long-tail keywords. Long-tail keywords are multiple words combined. Here are some examples:

- Canvas shoes
- Nerdy T-shirts
- Healthy chocolate
- Men's razor subscription

These are going to be much easier to rank for, and the more words you can come up with, the easier it will be to find a keyword you can try to optimize.

If I was just getting started with SEO, I'd try to make even longer-tail keywords, such as the following:

- Custom-printed canvas shoes
- Batman and Superman T-shirts
- Chocolate candy with health benefits
- Affordable men's razor subscription

The more specific your keyword choices are from the beginning, the quicker you'll see results. Once you've established a foothold, you can then aim for more competitive keywords.

Let's take an actual example from a local WooCommerce store, *The Chocolate Therapist* (`https://www.thechocolatetherapist.com`). They sell chocolate and talk about the health benefits (when eaten in moderation!).

Some keywords they might come up with include the following:

- Chocolate
- Dark chocolate
- Healthy chocolate
- Health benefits of chocolate
- Small batch of chocolate
- Gluten-free chocolate
- Dairy-free chocolate

- Soy-free chocolate
- Dairy chocolate
- Chocolate and wine pairing

You could go on and on and come up with lists of hundreds or even thousands of keywords.

Comparing search volume

Once you have your list, it's time to figure out which terms are the most popular. There are tons of tools that compare search volume on search engines. Some of the more popular tools are as follows:

- SEMrush (`https://www.semrush.com/`)
- Ahrefs (`https://ahrefs.com/`)
- Moz (`https://moz.com/`)

Many of these tools are paid. If you're going to invest in SEO, you'll want to get a premium SEO tool. However, if you're just experimenting with SEO, you can get an idea of search volume using Google Trends (`https://trends.google.com/trends`). This free tool allows you to enter different keyword phrases and compare their relative popularity. It won't tell you about the exact searches, as you'll need a paid tool for that, but you can get an idea of how widely searched certain terms are.

Here's my comparison of `healthy chocolate` versus `health benefits of chocolate`:

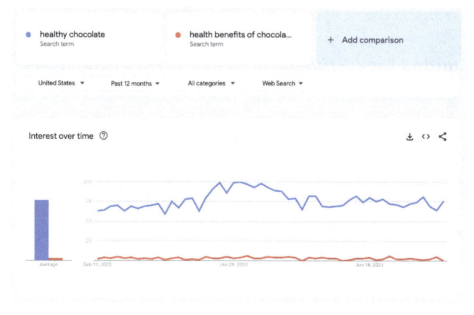

Figure 4.1 – Google Trends comparing "healthy chocolate" with "health benefits of chocolate"

As we can see, `healthy chocolate` is much more popular than `health benefits of chocolate`. By doing this, I know that I should use that phrase throughout my site and minimize using `health benefits of chocolate`. Google Trends shows you a small number of keywords. Some of the paid tools let you upload a CSV file to compare hundreds or thousands of keywords.

We'll learn how to optimize the process of searching for keywords in the next section.

Optimizing for keywords

After identifying the most popular words and assessing keyword competition, you can incorporate this terminology across your site. You can use it in your product descriptions, blog posts, URLs, downloads, tag lines, page titles, and so on.

As an example, I might do research and find that *two-player games* are more popular than *games for couples*. In this case, I should weave *two-player games* into the body section, into blog posts, into image alt tags, and maybe even into product titles.

You don't want to spam your site. It should always sound natural if you read it out loud, but instead of using whatever sounds good at the time, you'll want to use a keyword so that search engines know what your site is about and will show it on the **search engine results page (SERP)**.

There are entire books written about SEO, but this is a quick overview that should help you get started with optimizing your site for search engines. As with most marketing strategies, there's always more you can do, but this should help you appear in search engines and get your first free visitors.

Next up is a strategy we can use to help us make sure we organize our site to bring in the traffic from search engines.

Configuring breadcrumbs for search engines and users

Let's say a visitor lands on the **caramel and dark chocolate** product category of the *The Chocolate Therapist* site. They might like what they see, but they might want to look at dark chocolate without caramel.

If you build your site poorly, the user will have to go back to the **Shop** or **Home** page and then navigate to the right category.

By default, WooCommerce will add breadcrumbs to your product and category pages. Breadcrumbs show your users how to navigate to a parent category or back to the **Shop** page.

In the following example, we can see the **Home | Clothing| Accessories | Hat** breadcrumbs. In Storefront, a product page will look something like this:

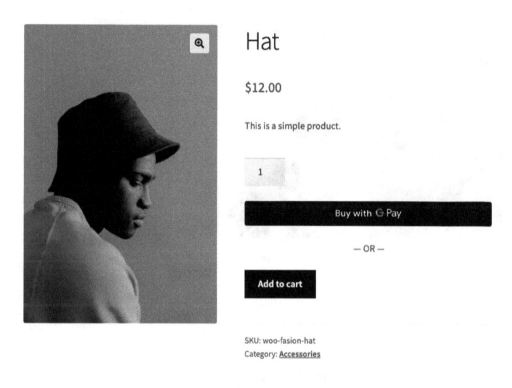

Figure 4.2 – Breadcrumbs on a product page

A category page will look something like this:

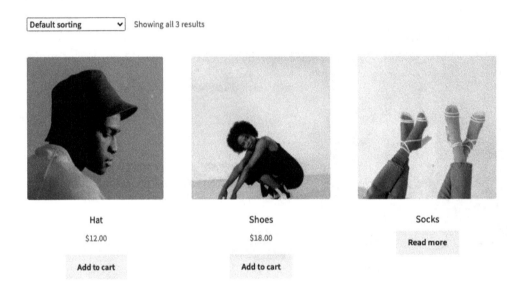

Figure 4.3 – Breadcrumbs on the category page

Breadcrumbs are controlled entirely by WooCommerce. However, they can be modified by your theme. By default, any theme should have breadcrumbs on the product and category pages. But if your theme does something atypical, you might not see them.

If you want to add or customize your breadcrumbs, the best way is to add a custom **personal home page (PHP)** code to your theme.

Adding custom PHP code

If you're familiar with customizing themes, you can add a little bit of code to your theme and breadcrumbs will automatically be added.

Add the following code to a theme template and you're good to go:

```php
<?php woocommerce_breadcrumb(); ?>
```

Alternatively, if you want to customize the functionality, there are a few parameters you can adjust. You can pass in a $args array and in that array, you can modify the following fields:

- delimiter: The character to display between the breadcrumbs
- wrap_before: The breadcrumbs container's starting code
- wrap_after: The breadcrumbs container's ending code
- before: HTML to display before the breadcrumbs
- after: HTML to display after the breadcrumbs
- home: Include the front page at the beginning of the breadcrumbs

Storefront and other themes are generally coded in a way where you don't have to add code to add the breadcrumbs. If your theme already has breadcrumbs, you can customize them with a filter.

Here's an example where I customized delimiter (what is displayed between the links), and I manually added the fries emoji (🍟):

```php
<?php
add_filter( 'woocommerce_breadcrumb_defaults', ' mastering_woo_change_
breadcrumb_delimiter', 20 );

function mastering_woo_change_breadcrumb_delimiter( $defaults ) {
// Change the breadcrumb delimeter from '/' to '🍟'
$defaults['delimiter'] = ' &#127839; ';
return $defaults;
}
```

This is a bit of a silly example, but it shows that you can add *any* character as the delimiter. However, you do need to represent the character as an HTML entity (https://www.w3schools.com/html/html_entities.asp). That means using 🍟 instead of 🍟.

If you want to customize the breadcrumb, you can refer to the official documentation (https://woocommerce.com/document/customise-the-woocommerce-breadcrumb/).

Using a plugin

Using plugins is another way to customize the breadcrumbs on your store. One of the more popular plugins is *WooCommerce Breadcrumbs* (`https://wordpress.org/plugins/woocommerce-breadcrumbs/`); it's shown in *Figure 4.4*:

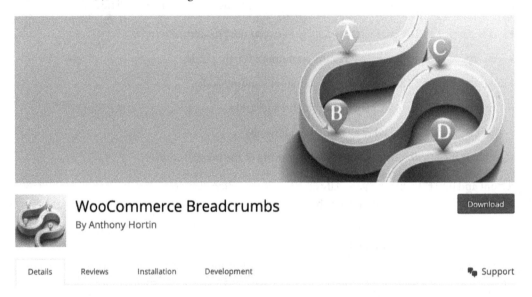

Figure 4.4 – The WooCommerce Breadcrumbs plugin on WordPress.org

This plugin is easy to use, and it does what you need without having to dip into the code.

Now that we've made our site easier to navigate, let's make sure search robots know exactly what is on our site with an XML sitemap.

Creating and sharing an XML sitemap

An important tool to help search robots such as Google understand the structure of your site is an **XML sitemap**. An XML sitemap lists a website's important pages, ensuring that Google and other search engines can locate and index them, thereby enhancing their comprehension of your website's structure

Here are the first few listings on the sitemap I use on my site:

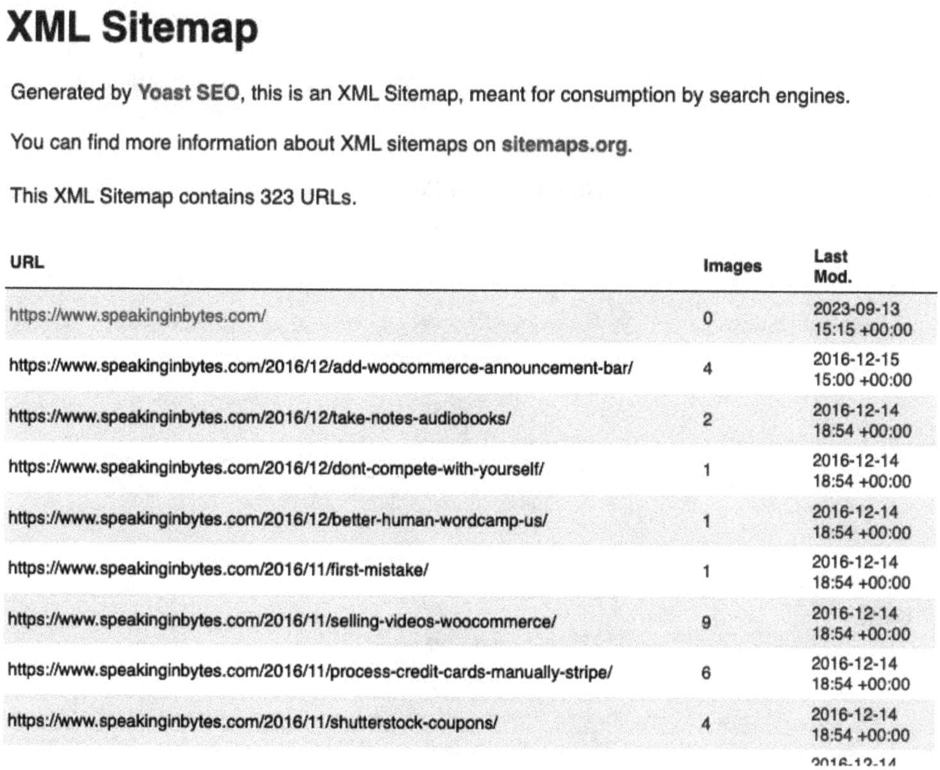

XML Sitemap

Generated by **Yoast SEO**, this is an XML Sitemap, meant for consumption by search engines.

You can find more information about XML sitemaps on **sitemaps.org**.

This XML Sitemap contains 323 URLs.

URL	Images	Last Mod.
https://www.speakinginbytes.com/	0	2023-09-13 15:15 +00:00
https://www.speakinginbytes.com/2016/12/add-woocommerce-announcement-bar/	4	2016-12-15 15:00 +00:00
https://www.speakinginbytes.com/2016/12/take-notes-audiobooks/	2	2016-12-14 18:54 +00:00
https://www.speakinginbytes.com/2016/12/dont-compete-with-yourself/	1	2016-12-14 18:54 +00:00
https://www.speakinginbytes.com/2016/12/better-human-wordcamp-us/	1	2016-12-14 18:54 +00:00
https://www.speakinginbytes.com/2016/11/first-mistake/	1	2016-12-14 18:54 +00:00
https://www.speakinginbytes.com/2016/11/selling-videos-woocommerce/	9	2016-12-14 18:54 +00:00
https://www.speakinginbytes.com/2016/11/process-credit-cards-manually-stripe/	6	2016-12-14 18:54 +00:00
https://www.speakinginbytes.com/2016/11/shutterstock-coupons/	4	2016-12-14 18:54 +00:00
		2016-12-14

Figure 4.5 – The XML sitemap for my site

As you can see, I have 322 URLs on my site. The sitemaps list them all and make it easy for search engines to understand when they were last updated, which category they're in, and where they're located.

> **XML sitemaps are now in WordPress core**
>
> One of the incredible things about WordPress is that it's always evolving. In the first iteration of this book, you had to install an SEO plugin to configure your XML sitemaps.
>
> As of WordPress 5.5 XML, sitemaps are included in WordPress core (`https://make.wordpress.org/core/2020/07/22/new-xml-sitemaps-functionality-in-wordpress-5-5/`).
>
> While WordPress itself includes XML sitemaps, you should still submit your XML sitemap to Google so that it knows that there is a sitemap and where to find it.

Submitting an XML sitemap to Google

Search engines can find your XML sitemap without you having to do anything. However, to make sure the XML sitemap is utilized fully, you should submit it to Google Search Console (formerly Google Webmaster Tools: `https://search.google.com/search-console`). Follow these steps:

1. Head over to **Google Search Console** and create a new site.

2. Once you've created your site in **Google Search Console**, you'll see a link for **Sitemaps**. Here's what **Google Search Console** looks like for my site:

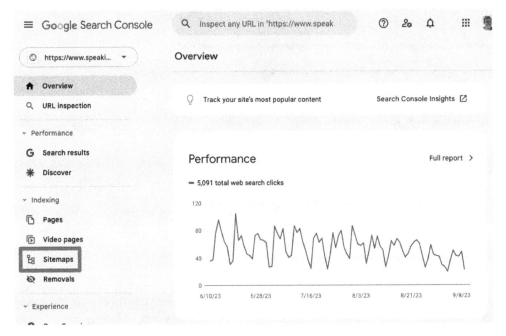

Figure 4.6 – Sitemaps in Google Search Console

3. From here, you just type in the URL of your sitemap that's generated by WordPress. Typically, it's `sitemap_index.xml`:

Figure 4.7 – Adding the URL of your sitemap

4. Once you submit it, you should see it listed on that page. After leaving a little time for scanning, Google should show you how many links it discovered:

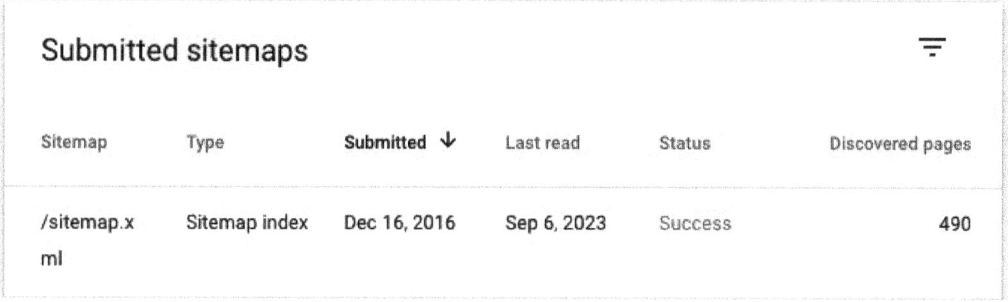

Submitted sitemaps

Sitemap	Type	Submitted ↓	Last read	Status	Discovered pages
/sitemap.x ml	Sitemap index	Dec 16, 2016	Sep 6, 2023	Success	490

Figure 4.8 – Submitted sitemaps in Google Search Console

With sitemaps submitted, our content and products will be indexed. When someone searches on Google, our site should be discoverable. Wait a few days and then do a site search, `site:yourwebsite.com`, on Google. You should see your pages in the search results.

If you don't see your site in the search results, something is wrong and you should check Google Search Console. This is what we'll cover next.

Keeping an eye on Google Search Console

I like to think of Google Search Console as an early warning system for my website. If something goes wrong and Google can no longer access your website or a page within your website, they'll let you know.

They sometimes send emails to the site administrator while other times, they'll have notifications within Google Search Console.

If you can log into Google Search Console once a month to make sure your website is ranking properly, that will help with SEO.

Here are two examples where I noticed a problem early and fixed it:

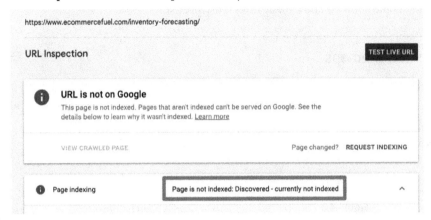

Figure 4.9 – Google Search Console flagged up an incorrectly indexed post

In this first example, Google noticed I published a new page but for some reason, it wasn't added to the index. That means it's impossible to find through the search engine. That's bad. Doubly so since I spent a month writing this piece!

The solution was simple. I clicked the **Request Indexing** link and within 48 hours, the page was reindexed, making it possible to find it through Google.

The second example is getting a notification from Google Search Console:

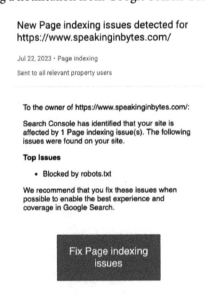

Figure 4.10 – Notification from Google Search Console

Sometimes, a plugin can misconfigure an important document, such as `robots.txt`. This tells search engines which pages they're allowed to index. You might accidentally turn off an entire category, posts, or something similar. Google Search Console can help you by sending you alerts when they think you're hiding too much content.

The solution is pretty straightforward: click the **Fix Page indexing issues** button and follow the steps provided.

Google Search Console is the tool that tells you when something is misconfigured with your site and Google. I know plenty of businesses that ignore Google Search Console and then wonder why their site won't show up at all in search engines. I always recommend that clients register their sites with Google Search Console and make sure someone is paying attention to their emails so that they can fix issues quickly.

If you're intentional with the keywords you use on your site, you use your keywords in your product descriptions and throughout your site, and you have breadcrumbs and an XML sitemap, you should be well on your way to having a website optimized for SEO.

Summary

In this chapter, we learned why SEO is so important, including how to find the right keywords, how to configure breadcrumbs, how to create and submit an XML sitemap for our sitemap, and finally why we should pay attention to Google Search Console.

Having optimized many parts of the store, we should see an increase in organic traffic from search engines such as Google. That will help our store stay in business and, hopefully, make a ton of money so that we can keep developing functionality for the store.

In the next chapter, we will learn how to manage sales in WooCommerce.

Part 2: Managing an Online Store

Once you have a store up and running it's time to start fulfilling those orders. In *Part 2*, you will learn to find your way around the admin to manage your store effectively. In this section, we will cover the following chapters:

- *Chapter 5, Managing Sales Through WP Admin*

- *Chapter 6, Syncing Product Data*

- *Chapter 7, Configuring In-Store POS Solutions*

- *Chapter 8, Using Fulfillment Software*

- *Chapter 9, Speeding Up Your Store*

5
Managing Sales Through WP Admin

Now that we've organized our products and set up our store to attract organic traffic from search engines, we need to think about how we fulfill orders. If your store takes off, you'll likely spend a lot of time processing orders, handling returns, and reviewing revenue and sales trends.

Processing orders is working *in* your business. It's what you need to do to get good reviews and get the next order. If you can process orders and refunds quickly, that gives you time to work *on* your business. That means reviewing sales and revenue trends to understand where your business is making a profit and doubling down on what's working as well as learning where you're unprofitable and where you might need to divest or run a clearance sale to remove inventory that's not moving.

This chapter will cover the following topics, which are just a few of the responsibilities that store owners have to handle:

- Fulfilling orders
- Refunding orders and payments
- Viewing internal sales data
- Using third-party sales data

Once you know how to do all of these things, you should be able to manage your own online store.

Technical requirements

WooCommerce handles many responsibilities for store owners. However, WooCommerce can't ship your packages. To do that, we'll install a few plugins that connect our store to services to help us ship our products and review sales information:

- Earlier in this book, we installed **WooCommerce Shipping & Tax** (https://woocommerce. com/document/woocommerce-shipping-and-tax/). We're going to fulfill some orders in this chapter, so, you'll need that installed.

- You'll also need **Jetpack** (`https://wordpress.org/plugins/jetpack/`) installed since it's a requirement for WooCommerce Shipping & Tax.

- Near the end of this chapter, we're going to look into some third-party tools including **Metorik** (`https://metorik.com/`), which has excellent reporting for online stores. They have a free trial, so you can install and sign up for the service.

Fulfilling orders

When people take out their credit cards to pay for things with their hard-earned cash, they expect you to ship their order quickly and safely. Getting a package from your warehouse to a customer's home is called **fulfillment**, and there are a bunch of smaller steps included. They are as follows:

1. Receiving a new order notification
2. Viewing customer shipping information
3. Picking items in the warehouse
4. Packing items into boxes
5. Printing shipping labels
6. Dropping off packages
7. Marking orders as complete

Exploring new order notifications

The whole process starts with a notification, which WooCommerce has automatically enabled. You can view the notification email under **WooCommerce | Settings | Emails**, as shown in the following image:

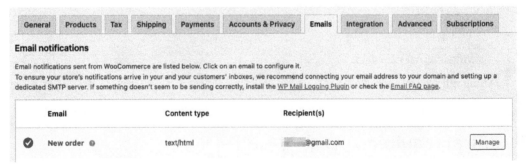

Figure 5.1 – Control WooCommerce email notifications

By default, these emails will be sent to the site admin. You can click **Manage** to change the recipient or add extra recipients.

New order badge in the site admin

If you're browsing through the WordPress admin panel, you can also see orders on the menu. Click on **WooCommerce** in the **Admin** menu, and you can see the number of new orders.

It will look like the following image:

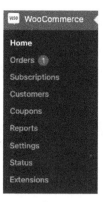

Figure 5.2 – New order badge in the WordPress admin

This shows the number of new orders. If you view the orders but you don't fulfill them, this notification bubble will disappear.

Browsing orders

If you want to see all of your orders, you can browse through them by clicking on **WooCommerce |** **Orders** from within the WordPress admin, as shown in the following image:

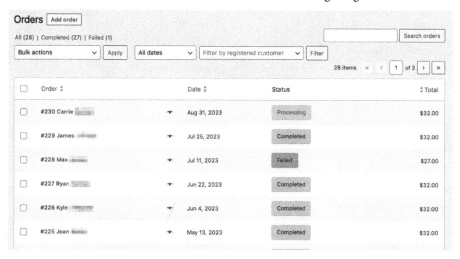

Figure 5.3 – Browsing orders in WooCommerce

You can very easily see which orders have a **Processing** status. *Processing* means an order has been paid for, but not yet shipped. Once you ship the order, you should change the status to **Completed**.

Either way, once you know that there's a new order, you now need to know what items need to be packed and where to send those items.

Viewing shipping information

On the **Orders** page in WooCommerce, you can click on the little eye icon next to the order as shown in the previous image. It will show you a little preview with the shipping details and a list of the items that need packing.

The details will resemble the ones in the following image:

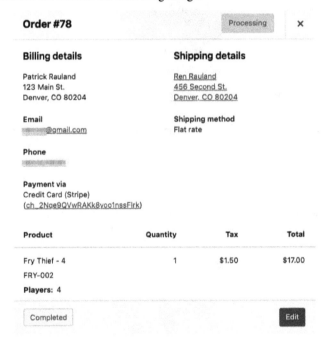

Figure 5.4 – The order preview in WooCommerce

> **Note**
> You may not see **Shipping details** in the order preview if you skipped the configuration of the shipping options in *Chapter 1*.

You could pack all of the items at the bottom of this screen, drop the package off at a shipper (courier service), and mark the order as complete by clicking **Completed**.

On the **Orders** page, you could also click on the **Order number | {{customer name}}** to see the **Edit order** screen with a lot more detail, as shown here:

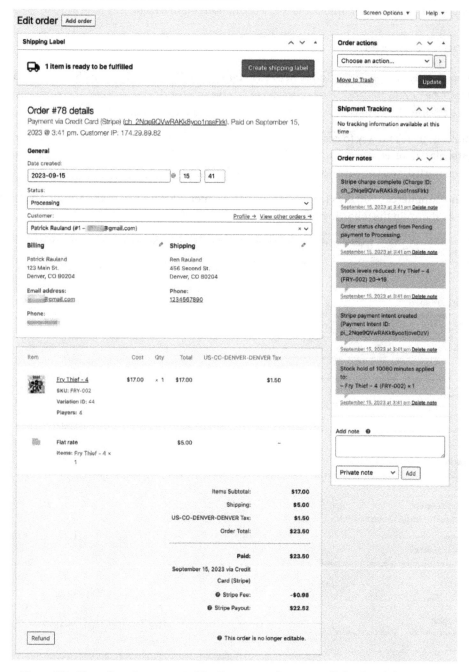

Figure 5.5 – The edit order screen in WooCommerce shows you all the details for an order

Here, we can again see shipping information, communicate with the customer, edit any information, print shipping labels, or initiate a refund. Loading a new page may take a bit more time compared to the preview feature, but you'll unlock the full set of options on this page.

Packing the boxes

Live shipping rates such as UPS, **United States Postal Service (USPS)**, and FedEx have a box packer built in (`https://docs.woocommerce.com/document/understanding-box-packing-calculations/`), which is how you can get accurate estimates on shipping costs.

Unfortunately, there's no way to access this box packer to see exactly which boxes the box packer uses for its estimate. However, you can select which boxes your box packer can use in the first place. If you're just getting started, you can save yourself a lot of time by having a small set of boxes that you use.

Go to **WooCommerce | Settings | Shipping | WooCommerce Shipping**. At the bottom, click **Add package**, as shown in the following image:

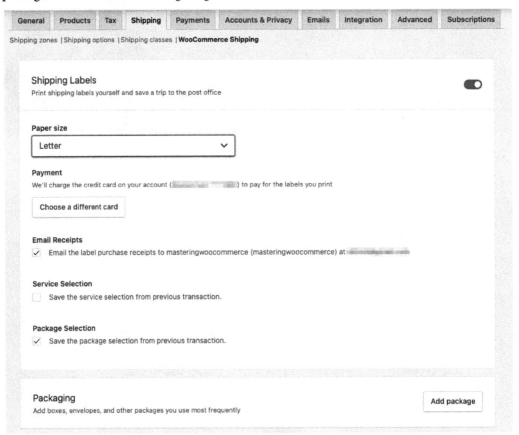

Figure 5.6 – Customize your shipping labels and packaging

> **Note**
> You have to have WooCommerce Services installed in order to see this menu option. If you have installed other shipping extensions, you might have to find similar settings under a different menu.

You can select pre-created boxes in the **Service package** tab, as shown here:

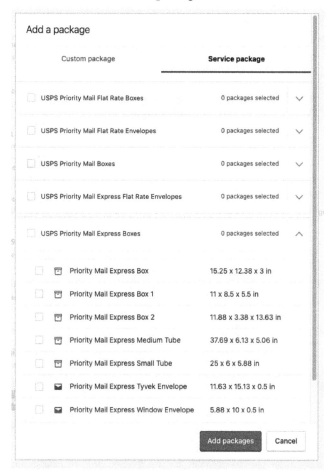

Figure 5.7 – You can select standard box sizes

You can also create brand new packages under **Custom package**. For most stores, you won't need to know how to do that. However, sometimes, you can save a lot of money if you tend to ship unique package sizes.

A good example of this is a sunglasses store, with which I have had a recent experience. They often sell one pair of sunglasses at a time, so they ship them in a very small, tight box, in order to protect the sunglasses and save shipping costs.

Printing shipping labels

Once you have your boxes ready to go, you can print out shipping labels. Using the **Edit order** screen, you have the opportunity to print out a shipping label, as seen here:

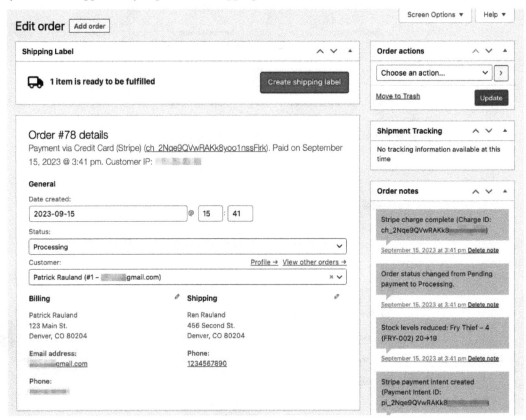

Figure 5.8 – Create a shipping label on the Edit order page

> **Note**
>
> This is powered by WooCommerce Shipping & Tax, so you have to have this plugin installed and activated to see this button.

You'll be asked to confirm the sender's and the receiver's address in the **Origin address** and the **Destination address** areas respectively. Then, you can select packages and add a weight (if necessary), as shown here:

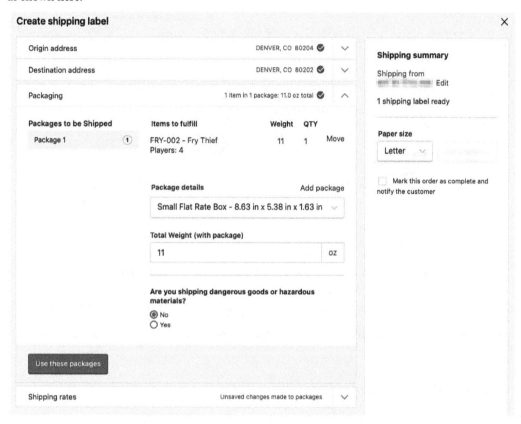

Figure 5.9 – Confirm Origin address and Destination address, then choose your packages

Click **Use these packages**, and then you can select your rate, as shown in the following image:

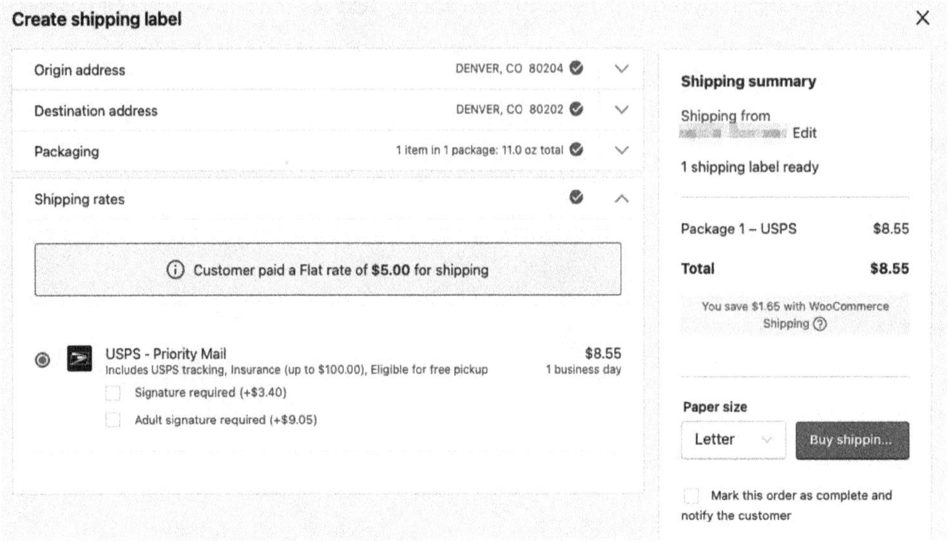

Figure 5.10 – Choose a rate and buy a shipping label in your admin

Finally, you can select **Buy shippin**. You'll get a PDF file, and you can print it out on your printer at home.

If you happen to have a label printer, you can change the **Paper size** field, and select a smaller size that will work with a label printer.

Dropping off packages

Once your packages are packed and have shipping labels attached, you're ready to drop them off. You can, of course, manually go into your shipping provider's office and drop off the packages.

However, once you start getting more than a couple of orders a week, you'll want to schedule pickups. Most shipping providers will let you schedule pickups ahead of time, and you can even have a recurring pickup time.

> **Note**
>
> If you're in the US, you can use USPS, which has a very easy-to-use tool to schedule a pickup. I won't go into detail here, but I've documented this process on my site (`https://www.speakinginbytes.com/2017/11/schedule-usps-pickup/`).

Marking orders as complete

Whether you print out labels or manually drop packages off in your own car, the last step is to mark an order as completed.

You can do this by checking the **Mark this order as complete and notify the customer** checkbox while purchasing your label, as seen in the previous screenshot.

You can also manually change the status of the order to **Complete** on the **Edit orders** page, as seen here:

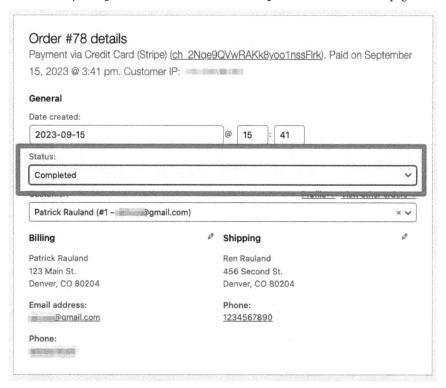

Figure 5.11 – You can always manually change an order status

Once you have marked an order as being complete, you're done. You can ship the rest of the orders and then move on to another task. Now, let's learn how to reimburse the users in the next section.

Refunding orders and payments

Of course, no matter how well you provide your services, such as advertising, describing, and shipping the products, some users will be unhappy and will want a refund. Refund rates vary wildly from industry to industry. Some industries will have refund rates in the low single digits, for example, 1-3%, whereas other industries, such as fashion, can have refund rates up to 25-30%, which is massive.

Once you've had a few refund requests, you'll want to build a process to make it easier. Let's start with manual refunds, and then look into building a refund process.

Refund requests

If a customer emails you, tweets you, or calls you for a refund, you'll have to know how to make a refund. This can be done on the **Edit order** page.

The following image shows the functionality:

Figure 5.12 – Start the refund process on the Edit order screen

You can click the **Refund** button, which changes the **Qty** field to an input field, where you can choose how many items you want to refund. In our case, let's say someone wants to keep only one hat, and they want to return the other one.

You can manually type in the reason (so it's logged) in the text box, as shown in the following image:

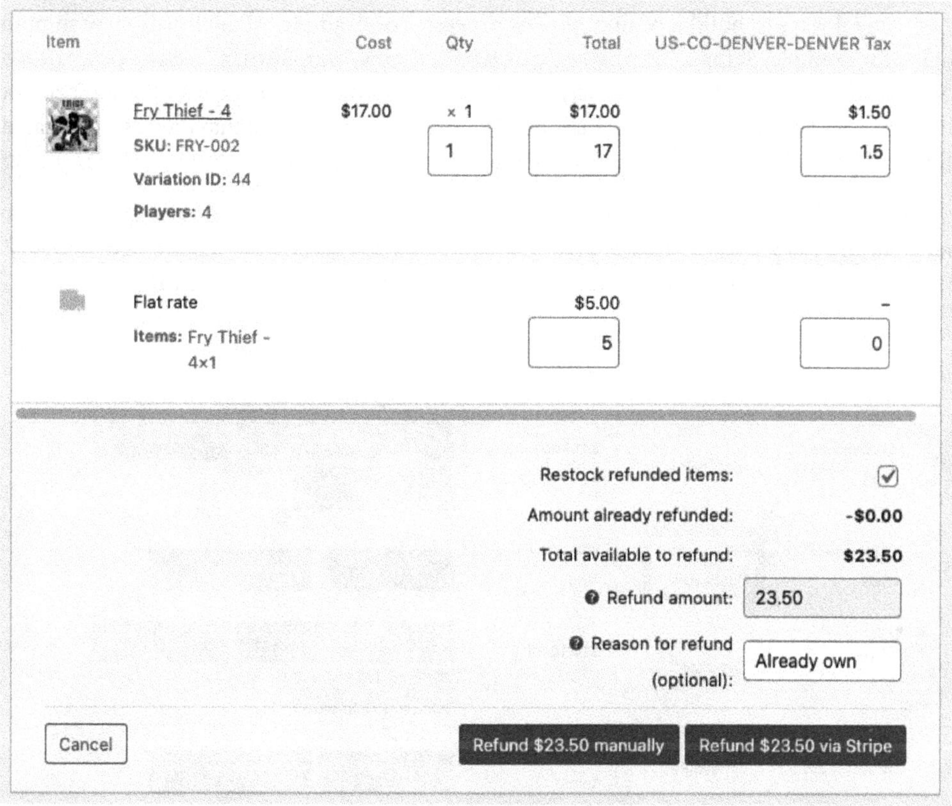

Item		Cost	Qty	Total	US-CO-DENVER-DENVER Tax
	Fry Thief - 4	$17.00	× 1	$17.00	$1.50
	SKU: FRY-002		1	17	1.5
	Variation ID: 44				
	Players: 4				
	Flat rate			$5.00	–
	Items: Fry Thief – 4×1			5	0

Restock refunded items: ☑

Amount already refunded: -$0.00

Total available to refund: $23.50

❷ Refund amount: 23.50

❷ Reason for refund (optional): Already own

Cancel   Refund $23.50 manually   Refund $23.50 via Stripe

Figure 5.13 – You can refund products, shipping costs, or a custom amount

Then, depending on the payment gateway that you used to pay for the order, you'll see two buttons to refund.

If you see the button on our screen that says **Refund $XX manually**, the order will be refunded in WooCommerce, but you still have to manually go into a payment gateway and refund their money in that system.

Some payment gateways, such as Stripe, support automatic refunds, and you'll see a **Refund $XX via {{Payment Gateway}}** button. If you click that, then the order will be refunded in WooCommerce, and it will also be refunded through Stripe.

If your payment gateway supports automatic refunds, you will almost always want to use that functionality so that you don't forget to manually refund in another system.

Building a refund process

There are several ways to build a refund request process. For example, many third-party shipping companies, for example, Shippo (`https://goshippo.com/products/post-purchase-tracking/`), have a refund request process built in. That lets customers request refunds, print out shipping labels, and send them in. They built this in a self-serve manner so that the customer can do all of this, and you only pay for shipping labels when they're actually sent.

However, let's say that you want to keep this in WooCommerce. You can use **WooCommerce Smart Refunder** (`https://woocommerce.com/products/woocommerce-smart-refunder/`), which builds a refund request button onto the automatically generated **My Account** page:

Figure 5.14 – Users can initiate their own refund from their dashboard

Then, customers can fill out a form to request a refund, as shown here:

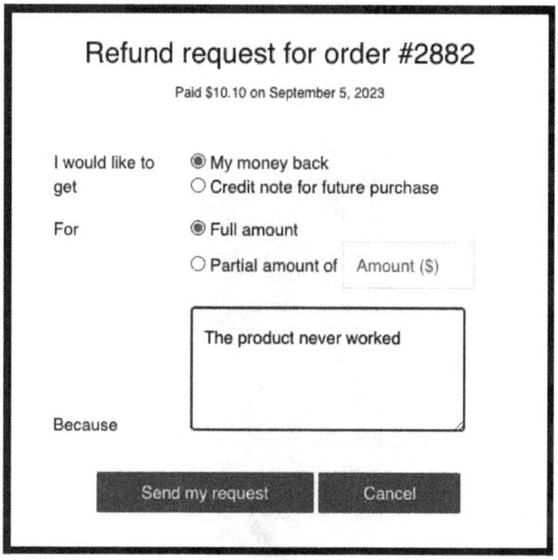

Figure 5.15 – Submit the refund request

The admin can build rules to manually or automatically refund certain types of orders. From the page where you moderate refund requests, if you approve the requests, they'll automatically refund the money and change the order status.

Having a process to refund orders is very useful, but it doesn't need to be in place when you launch your store. You can wait until you get a few refunds to see what you want your process to look like.

You should now have a system to refund any orders that come in, which is an essential feature for an online store. Now that you can fulfill and refund orders, you can dig into the order data.

Viewing sales data

Once you've shipped your orders, you'll probably want to see which products and categories that the customers are interested in. When you have this data, you can buy more of the most profitable and fastest-moving products. Luckily for us, WooCommerce recently improved this experience quite a bit!

Since the first version of this book, WooCommerce revamped their admin experience. They started the project with WooCommerce 4.0 (`https://developer.woocommerce.com/2020/03/10/woocommerce-4-0-is-here/`) and it was fully merged with WooCommerce 6.5 (`https://developer.woocommerce.com/2022/05/10/woocommerce-6-5-released/`). That's over two years of work of completely updating the admin experience and reporting within WooCommerce. If you tried WooCommerce reports before 4.0, it's worth taking another look at the new and improved reports in WooCommerce.

We're going to look into some basic data, which you can see at a glance on the **Orders** screen, as well as all of the data that you can dig into through the WooCommerce reports.

WooCommerce analytics

Okay, so you're excited to find their new reports feature. Where would you click in the following image?

Figure 5.16 – Admin menu in WordPress

Many of you probably guessed **Reports**, but those are actually the *old* reports. They still work but you'll want to access the new functionality. Click **Analytics**.

When you click **Analytics**, you'll see a ton of reports, which are generated by the new WooCommerce admin. There are several reports available to you. They are as follows:

- **Products**
- **Revenue**
- **Orders**
- **Variations**
- **Categories**
- **Coupons**

- **Taxes**

- **Downloads**

- **Stock**

- **Customers**

If you install additional extensions, such as Product Bundles, you'll see additional reports.

Out of all of these, **Revenue**, **Products**, **Categories**, **Taxes**, and **Customers** are my favorites. Let's take a look at the **Revenue** report to really dig in. It is shown in the following image:

Figure 5.17 – The Revenue report in WooCommerce Analytics

The first element that you notice is the pretty graph. The graph does a fine job, but the most useful tool on this page is the **Date range** feature. You can click on this and compare time periods. In most cases, you will want to compare against a previous month or a previous year.

This is my favorite way to look at all stats. Compare it against a previous time period, and make sure that the graph is going up and to the right. If so, you're doing great. Don't worry about absolute numbers:

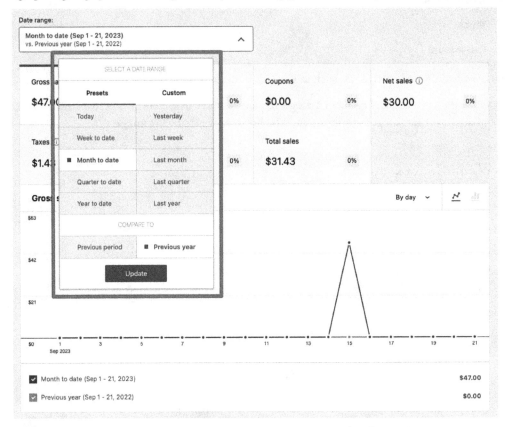

Figure 5.18 – Compare your results against a previous period or the previous year

These reports help us understand the following important metrics of our business:

- What products are selling
- Gross profit (revenue minus cost of goods sold)
- Taxes

Let's take a look at them, one by one, in the following sections.

What sells

The next two useful graphs are for **Products & Categories** (pictured in the next diagram). These are really good at showing you what is selling. You'll want to get more products such as those in order to keep up a steady flow of purchases. Here's an example of my category report:

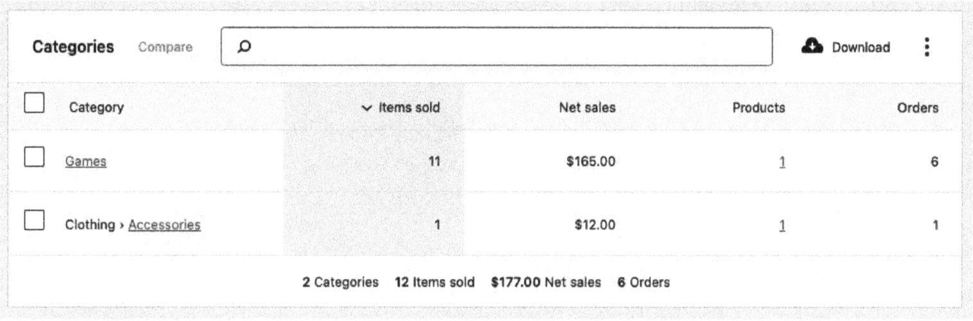

Category	Items sold	Net sales	Products	Orders
Games	11	$165.00	1	6
Clothing › Accessories	1	$12.00	1	1

2 Categories   12 Items sold $177.00 Net sales   6 Orders

Figure 5.19 – Category report shows which categories are performing

From my example, it seems that clothing accessories, such as hats, sell much worse than games. Now, I have to decide whether I want to promote games more and cross-sell my merch or if I should perhaps try advertising merch itself to see if that can sell.

> Tip
>
> One of the best ways to grow your business is to increase the **average order value** (AOV). Once someone is going to buy one product from you, it's not that much more work to convince them to buy another product. Think about the snacks in the grocery store. You're already ready to buy $100 of groceries and have your credit card ready; it's easy to add a $3 candy bar to the checkout.
>
> Porting this to the realm of e-commerce – I love helping businesses grow by adding accessories to their product line. If most of your customers are only buying a single product, then see if you can cross-promote another product or, ideally, an accessory that enhances the first product.

Gross profit

One thing that this report doesn't do is show you how profitable an item is. So, if you are selling hundreds of products that have a $1 margin, and you sell two items that have a $100 margin, you might actually make more money from the products that you sell fewer of because they have a much higher margin.

This isn't a book on calculating gross profit, but these reports are a great start. You can export to third-party tools and import the cost of goods numbers. Or, you can look into the **WooCommerce Cost of Goods** extension (https://woocommerce.com/products/woocommerce-cost-of-goods/) to help you with these detail-heavy calculations.

> **Tip**
>
> **Margin** is the term for how much money you make on a product after subtracting the cost of the product itself. If you sell a product for $20 that you bought for $5, you have a $15 margin.
>
> You can calculate your profit margin with the following formula:
>
> Gross profit margin = (revenue – cost of goods sold / revenue) x 100
>
> In our case, we have a 75% profit margin. Generally speaking, that's a very good number.

If you have discount programs, product bundles, loyalty programs, advertising costs, and so on, they can all be factored into the preceding formula. When you're working with marketing agencies and the like, you can give them a gross profit target, so they know they can't give away too much profit to attract a sale.

Taxes

The other report that you'll definitely want to familiarize yourself with is the Taxes report. This shows you all of the taxes that you have collected, divided by type. So, if you collect a state tax, municipality tax, and a county-level tax, all of them will be listed separately, as shown in the following image, you know exactly where each tax should go:

Tax code	Rate	Total tax	Order tax	Shipping tax	Orders
US-CO-US-CO-DENVER-DENVER TAX-1	8.81%	$12.95	$12.95	$0.00	5
US-CO-US-CO-ADAMS-WESTMINSTER TAX-1	4.75%	$1.43	$1.43	$0.00	1
2 tax codes		$14.38 total tax	$14.38 order tax	$0.00 shipping tax	6 orders

Figure 5.20 – The Taxes report shows how much tax you owe and to which authority

There are, of course, some third-party tools that do similar things, but it's nice to have a good tax report feature built into WooCommerce. Even if you do use a third-party solution, you can always double-check their numbers with the built-in tool.

Using third-party reporting platforms

There are many services that help with reporting. There are services, such as Google Analytics, that help to measure traffic, sessions, and what content users are consuming on your site, and there are services, such as Metorik, that add e-commerce data on top of traffic data. We're going to look into Metorik since it was originally built to support WooCommerce store owners.

Exploring Metorik

Metorik (`https://metorik.com/`) is built for WooCommerce, and they have very detailed reports that you won't find anywhere else. If you really want to dig into the data of your store in order to make data-informed decisions, Metorik gives you some of the most comprehensive data and visualizations.

They have a free trial, so feel free to explore the service. We can see the dashboard in the next image, and it already gives us tons of information:

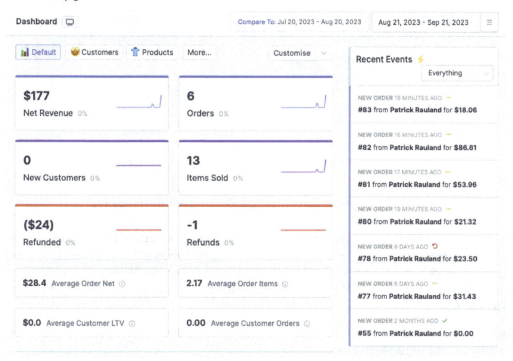

Figure 5.21 – The dashboard from the third-party reporting platform Metorik

When you log in, you'll see a dashboard with important numbers, often called **key performance indicators (KPIs)**.

In addition to the standard reports, such as Revenue, Orders, and Refunds, Metorik has a predictive algorithm. They added a magic wand icon above almost all of their charts (`https://metorik.com/blog/forecast-any-chart-with-just-a-click`). You can click this icon and it will forecast your future data and add it to the chart, as shown here:

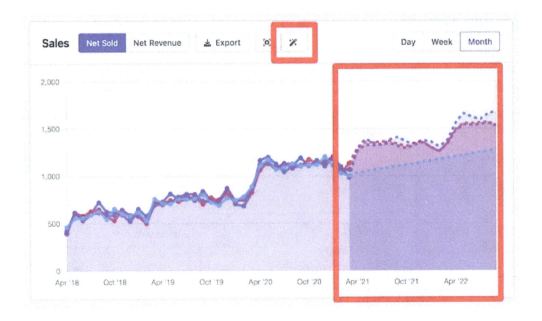

Figure 5.22 – The forecast feature built into Metorik charts is very powerful

This also applies to subscriptions (`https://metorik.com/features/subscriptions`). This data is incredibly important. It can calculate for how long customers will be customers (retention rate), as well as when they'll likely stop paying (churn). When you combine these numbers with your subscription price, you can calculate your customer lifetime value, which is basically how much money you'll make from a customer.

When you know your customer lifetime value, you know the absolute maximum that you can spend on a customer and still make money.

Pick one

There are a lot of great tools out there. You don't need to subscribe to all of them. Pick one, or maybe two tools that give you useful numbers, and stick with them.

Since Google Analytics is the standard tool and it's free, I recommend that you set it up, even if you don't plan on using it. A couple of years from now, if you change your mind, or you have a contractor that needs the tool, you'll be glad you have the data spanning across several years.

Summary

Store owners have a lot to do. Some of the most important duties are fulfilling and refunding orders. You might have to do both of these daily if your store takes off.

However, store owners will also have to review sales data both inside and outside of WooCommerce so that they know which products are selling well, as well as how much their customers are spending and how long they're customers for.

Now that you know how to do all of these things, you can manage your own online store. In the next chapter, we're going to look into growing your store and syncing data between different storefronts.

6

Syncing Product Data

We now know how to fulfill orders and manage sales for our store. The next important step is to list our products where our potential customers will see them.

When you run an e-commerce business, it's tempting to list your products on as many marketplaces as possible so that you can earn as much revenue as possible. A great example of this going wrong is when you have a WooCommerce store where you list a kite for $10, and then you have an eBay store where you sell the same kite for the same price. If you update one price and forget to update the other, you can run into the following problems:

- You could miss out on additional income
- Customers could find out they paid a higher price and demand a refund (or, even worse, they refund on one system and rebuy the exact item on your other site)
- You could oversell a product and not have enough stock to fulfill orders

In general, it's a bad idea to store data in two or more places. You always want to store it in just one place. This principle is called the **singular source of truth** (**SSOT**) (https://en.wikipedia.org/wiki/Single_source_of_truth). The goal is to only ever list a piece of data (such as a product price) in one place and then have every other system talk to your singular source of truth. This prevents a ton of issues where data is outdated.

In this chapter, we're going to look at the following three practices you can use to sync data:

- Exporting data out of WooCommerce
- Importing data into WooCommerce
- Integrating with an **Enterprise Resource Planning** (**ERP**) system

By the end of this chapter, you will be able to make sure your data is stored correctly so that you can avoid syncing and inventory problems, as well as learn how to import and export data and effectively use an ERP system.

Exporting out of WooCommerce

If you need to import data into another system, the first step is knowing how to correctly export that data out of WooCommerce. You can use WooCommerce as your singular source of truth, update all data in WooCommerce, and then export the data out of it into other systems.

If you need to copy data to another system, you should set up a direct integration, such as WooCommerce and Square integration (which we will set up in the next chapter), which automatically syncs products and product data between both systems.

However, if you don't have a direct integration, then you'll need to export data manually under **WooCommerce | Products | Export**, as follows:

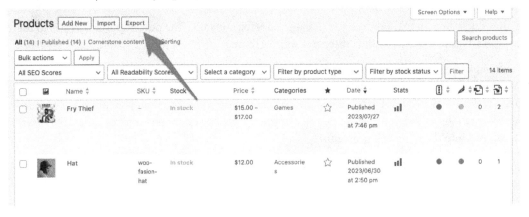

Figure 6.1: Exporting products from the Products screen

And you can import this data into different systems. We'll cover the ins and outs of a **comma-separated value (CSV)** file shortly.

Alternatively, you could use a tool such as Zapier (`https://zapier.com/`) that lets you connect your WooCommerce store with hundreds of other online applications. You can configure Zapier to watch for product updates and notify another system with each update.

If you use multiple systems to sell your products, you might have to export your product data manually and then load it into the other system. Now that you know how to export data, for the rest of this chapter, we're going to focus on syncing WooCommerce with these other systems. Let's see how to import products in the following section.

Exploring a CSV file

Before we go any further, it's worth explaining, briefly, what exactly a CSV file is. A CSV file is technically just text, and if you open it up with a text or code editor, you'll see something similar to this:

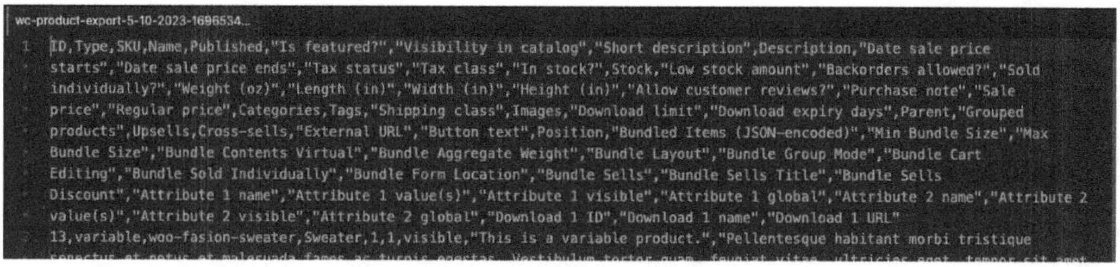

Figure 6.2: Raw CSV data

In its raw form, it's not very useful. However, text separated by commas is very easy for a computer to parse, which is why it's such a useful medium for sending data. If you open the same file in a program capable of displaying this information, you'll see something very different. You can see the same CSV file in the following manner in Google Sheets, as shown in the following screenshot:

Figure 6.3: CSV data in a spreadsheet

CSV files can very easily be opened by spreadsheet programs such as Microsoft Excel, Numbers, and Google Sheets, and you can easily export a CSV file from any of these programs.

> **Understanding delimiters**
>
> When you create a CSV file, you might be prompted to choose delimiters and a few other settings. In most cases, you should be able to use the defaults. A delimiter is a symbol between columns. In Europe, the comma character is used for a decimal – for example, €99,99. Because the comma is used in this fashion, some European software uses a semicolon as the delimiter in a CSV. If you're having trouble importing data from one system to another and you notice that columns aren't lining up, there's likely an issue with the delimiter, and you want to experiment with setting a specific character as the delimiter during both the import and export process.

Either way, WooCommerce lets you import any CSV file, and you can customize those settings on import to match your export.

Including content in a CSV file

Figuring out *exactly* what should be included in a CSV file is actually a complex process. If you try to do it manually, you are likely to make a mistake and have one too many or too few commas, which will mess up the entire export/import process.

To make this easy on yourself, start by exporting the default products in WooCommerce. You can even create a demo product or two, with settings similar to your products, and then export those.

You can do this under **WooCommerce | Products | Export**, as shown in *Figure 6.1*. The default options should be fine. They are shown in the following screenshot:

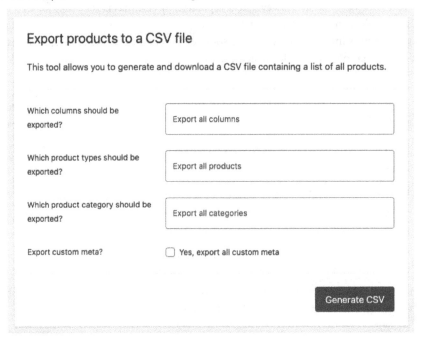

Figure 6.4: Exporting products out of WooCommerce

Click **Generate CSV**, and now, you should have a perfect template. You can delete the sample products and replace them with your own products. To make this process a little easier, if you have your own product spreadsheet, rename your columns to match the columns in the WooCommerce CSV export. That way, it's dead simple to match columns during the import process.

Importing products via CSV

One of my first WooCommerce projects was for a large furniture company that had its own proprietary database, where it kept all of its product information. This database was where it added new products, updated prices, added photos, and placed products into categories.

The company wanted its website to reflect what was in the database. So, we had two options:

- Integrate with WooCommerce so that every database change is mirrored on the website
- Export products into a CSV file and import it into WooCommerce

Functionally speaking, both approaches achieve the same results. However, having to maintain integration with an unfamiliar proprietary database sounds like a lot of work and a potential headache. Conversely, exporting products into a CSV file is easy, and it's equally easy to import that CSV into WooCommerce.

We're now going to look at how you can import a CSV file into your WooCommerce store.

Importing a CSV

From the previous section, we have a CSV file ready to go, so it's time to import it:

1. We can do this under **WooCommerce | Products | Import**, as shown in the following screenshot:

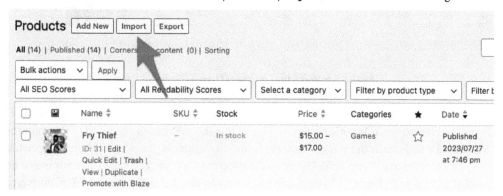

Figure 6.5: Importing WooCommerce products

Now, we get to go through the import process.

2. Upload your file and choose whether you want to update existing products. New products will be skipped. This is a useful option when we only have a small selection of products on our site. Otherwise, I recommend leaving this option disabled. You can see the settings for importing a file in *Figure 6.6*:

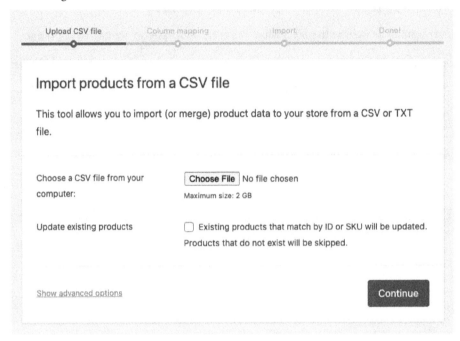

Figure 6.6: Importing products from a CSV file

3. The next step is the most important. Make sure that the data is imported into the right fields. If you used the exported WooCommerce CSV file, this should be pretty easy. All of the columns in the CSV will match the WooCommerce fields – for example, the Name column will match the WooCommerce Name field. If you used a different CSV file, the Name column might be called Title or Product Title, and you'll have to match it to the correct WooCommerce field, as you can see in *Figure 6.7*.

Double-check whether the columns in the CSV (left) match the values in WooCommerce (right). You can see the columns in *Figure 6.7*:

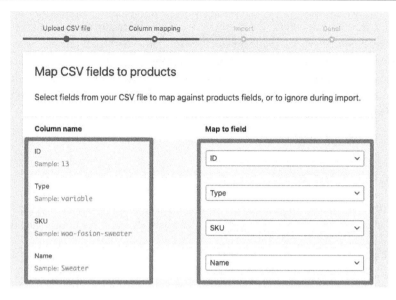

Figure 6.7: Match the column names with the WooCommerce fields

This might take a while, depending on how many products you're importing. If you link to a bunch of images in the CSV file, they will definitely take some time to import. Once you are done, you should see an **Import complete!** screen, as follows:

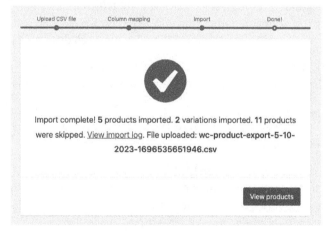

Figure 6.8: Import complete!

You can view the log and see whether any products were skipped or failed. If you want to update products, then you can go back to the beginning and click the checkbox to update existing products.

The trickiest part is making sure your data is mapped to the correct field in WooCommerce. If you can get that part correct, the rest of this process only takes a few minutes, and the import will go smoothly.

Now that you can import data via CSV into your WooCommerce store, you can keep all of your product data in sync, no matter what platform you use. You can export from one platform and import into another. This will make sure that product prices are consistent, out-of-stock products are listed as out of stock on all platforms, and users always read the most accurate and up-to-date product descriptions. We can build an online business with a singular source of truth that keeps us sane and our customers happy.

In the next section, we'll integrate with software that is a singular source of truth.

Integrating with an ERP

As your store grows, it's natural to slowly piece together different pieces of technology to solve the day-to-day challenges of running a business. You'll likely have different systems for the following:

- Accounting
- Taxes
- Shipping
- Inventory

This is fine for small stores, but as you grow, it becomes a lot to remember. For example, you might sell 10 bicycles a month and have 11 bikes left. A customer buys a bike, and WooCommerce sends you a low-stock notification. You ask your vendor for a quote/invoice and wait to hear back. You pay the invoice and log it in your accounting system. You receive the bicycles, and you have to mark the order as received. Then, you log the bicycles in WooCommerce so that new users can buy them.

This is a delicate process and breaks the single-source-of-truth principle we talked about earlier because, at certain points, you have to remember to log the same event in multiple systems.

Luckily, there is an easier way. You can use an ERP system. These are hubs for your entire business, and they integrate with many different tools, so you only have to enter something once and the data gets shared with your other tools.

Let's look at the earlier example again using an ERP – a customer buys a bike, and your ERP sends you a low-stock notification. You go to your vendor and ask for a quote/invoice. You log the paid invoice in the ERP. You receive the bicycles and mark the order as received. WooCommerce is automatically updated with the new quantity. Now, since everything is logged in one place, you have accurate numbers for the health of your business (taking into account revenue and costs).

Finding an ERP

There are hundreds of ERPs, and they all focus on different aspects of e-commerce. Some are really good at accounting, some at shipping, and some at managing customers and tracking all data related to them.

I tend to value the following two aspects:

- **Plug and Play integrations**: These include QuickBooks, ShipStation, Amazon, eBay, Walmart (and even other Shopify or Magento shopping carts), Square POS, and, of course, WooCommerce
- **Procurement tools**: Re-ordering, demand forecasting, and purchase orders

If you can sync product details across platforms automatically and re-order your products on time, it will really help you manage your cash flow, avoid stock-outs, reduce human errors, and, ultimately, make you more profitable.

I recently worked on a piece for eCommerceFuel with e-commerce veteran Patrick Mulligan. We created a list of enterprise-level, mid-level, and entry-level ERPs: `https://www.ecommercefuel.com/erp-ecommerce/`.

Here are the ERPs we covered:

- Enterprise-level ERPs:

  - NetSuite Oracle (`https://www.netsuite.com/portal/home.shtml`)
  - SAP Business One (`https://www.sap.com/products/erp/business-one.html`)
  - Microsoft Dynamics 365 (`https://dynamics.microsoft.com/en-us/`)

- Mid-level ERPs:

  - Brightpearl (`https://www.brightpearl.com/`)
  - Cin7 Core (formerly DEAR) (`https://www.cin7.com/`)
  - Fulfil.io (`https://www.fulfil.io/`)
  - Odoo (`https://www.odoo.com/`)
  - Zoho (`https://www.zoho.com/`)

- Entry-level ERPs:

  - Cin7 (`https://www.cin7.com/`)
  - Finale Inventory (`https://www.finaleinventory.com/`)
  - Shiphero (`https://shiphero.com/`)
  - Skubana (`https://www.skubana.com/features`)

As you can see, there's a huge variety! And these are just some of the more well-known solutions.

We're going to set up **Finale Inventory**, since it handles the two most important use case for us – inventory management and purchase orders. It also has a 14-day free trial, so feel free to set up an account and follow along.

Configuring Finale Inventory

The first thing you're going to have to do with any ERP is connect it to WooCommerce. In Finale Inventory, you can connect your store by going to **Integrations | WooCommerce** and clicking the **Add WooCommerce integration** button, which you can see in the following screenshot:

Figure 6.9: Adding WooCommerce integration in Finale Inventory

This will use the WooCommerce REST API. By default, the REST API should be enabled. If it's not, then you can enable it under **WooCommerce | Settings | Advanced | Legacy API**. Then, you can go to **WooCommerce | Settings | Advanced | REST API** and click **Add Key**.

Understanding the WooCommerce REST API

WooCommerce has a REST API. An **API** is an **application programming interface**. It's a way for software to talk to each other and share data. A RESTful API is a specific way to design an API. You can learn more about REST here: `https://en.wikipedia.org/wiki/REST`.

And you can dig into the details of the WooCommerce REST API here: `https://woocommerce.github.io/woocommerce-rest-api-docs/`.

The key takeaway is that a REST API allows WooCommerce to talk to an ERP.

The future of the WooCommerce REST API

The legacy WooCommerce REST API will in the not-too-distant future be moved to a separate plugin. It will still be available, but you'll have to download and install a separate plugin. You can read more here: `https://developer.woocommerce.com/2023/10/03/the-legacy-rest-api-will-move-to-a-dedicated-extension-in-woocommerce-9-0/`.

On the next screen, you'll be prompted to create a key for this integration. You can add a description, select a user to assign this key to, and give them permissions. For an ERP, you're very likely going to want to give them read and write permissions to update data on your site. See *Figure 6.10* to understand how I configured my key:

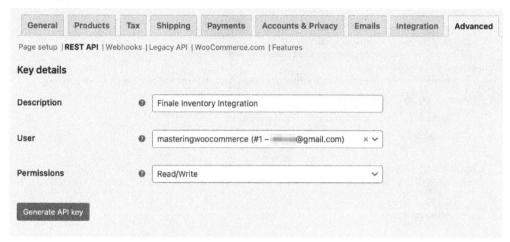

Figure 6.10: Creating a REST API key

Then, you'll see your **consumer key** and **consumer secret**. It's important to copy these down somewhere or copy them into Finale Inventory immediately. Once you exit this page, you won't be able to see this information again, and you'll have to create a new key.

Figure 6.11: The created REST API key

After copying the keys and the URL for your site into Finale Inventory, you can click **Test integration**.

1. Login to the admin screen for your WooCommerce store.
2. Select WooCommerce > Settings > Advance tab.
3. Click on the Legacy API tab, and enable checkbox for "Enable the legacy REST API"
4. Click on "Save changes" button
5. Click on the REST API tab, and go to Keys/Apps section by clicking on "Add Key" Button.
6. Note the Consumer Key and Consumer Secret Generated.
7. Fill the description field, select the user for whom keys needs to be generated, select level of access as 'Read/Write'
8. Click the "Generate API Key" button.
9. Enter the base URL for your WordPress installation, copy the Consumer Key, and copy the Consumer Secret in the fields below:

WordPress base URL:	https://masteringwoocommerce.mystagingwebsite.com/
Consumer Key:	ck_
Consumer Secret:	cs_

Success
Finale successfully connected.

Test integration

Figure 6.12: A successful connection test on Finale Inventory

When you see **Success**, make sure to click **Save Changes** at the top of the screen, and you're good to go. You can now use the ERP to synchronize all of your product details, inventory levels, purchase orders, and more.

ERPs are notorious for being hard to set up (and this is one of the easiest I found).

Importing products into ERP

This ERP makes it easy to import your products. Under **Inventory | Products**, click **Import from integration**.

Let's add products to your account

The best way to start using Finale Inventory is to add your products to the system. We provide several ways to do that, please use one of the following options.

Import products from an external system

Directly import products from external systems like stores, marketplaces or other integrations. We support a large number of external systems and add new ones frequently.

Import from integration

Figure 6.13: Importing products from an external system

You should see any existing connections, including the one we just set up. Here's what I see:

Figure 6.14: Importing products from WooCommerce into Finale Inventory

Click on **Import**.

On the next screen, you can choose how to perform the import. I definitely recommend importing stock. You can also start adding locations to Finale Inventory, such as multiple warehouses and retail locations. ERPs are great at storing all of this data.

Keep going through the wizard, and at the end, you'll be notified by email when the import is done. It only took a minute or two for my store.

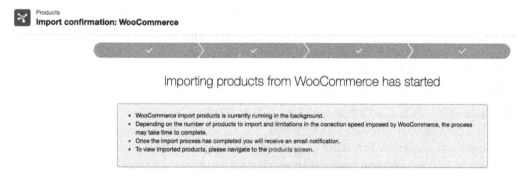

Figure 6.15: The ERP import starting

Once your products are imported, you can delve deeper into how to use an ERP.

Using an ERP

ERPs are very challenging to set up and require a lot of technical knowledge, but once the setup is complete, they are amazingly powerful tools. They also tend to be expensive. Many of these platforms cost over $100 a month.

I recommend setting one up as soon as you start entering lots of information into multiple systems. When it prevents you from making mistakes and saves hours of data entry, you'll see that it's worth the setup time and the cost involved.

Since ERPs take so long to set up, the switching costs are high – meaning you're very likely to stay with an ERP even if it doesn't do everything you want. For that reason, I recommend that you do the research and decide which ERP is right for your company.

There are dozens of well-designed ERP solutions, and we can't cover them all here. I made a list of several popular ERPs earlier in this chapter, and I have written about ERPs in the past. I also recommend reading through this guide to help you understand the pros and cons of various ERPs: `https://www.ecommercefuel.com/erp-ecommerce/`.

If you're importing and exporting CSVs frequently, eventually you're going to make a mistake and import data incorrectly. A direct integration with an ERP prevents those mistakes and can make your business run more smoothly.

Summary

If you only have one store, you shouldn't have to worry too much about syncing data between multiple systems. But as your business grows, you'll likely need to share data between systems.

In this chapter, we learned how to export and import CSV files from and into your WooCommerce site. That means you can connect *any system* to WooCommerce, and that is incredibly powerful. By importing and exporting the right data, you can test any piece of software and see whether it works for you and your business.

We also learned how to use an ERP system that can keep all of our data in one place. With an ERP in place, you don't need to sync data with CSV files, and preventing human error should help your business run more efficiently. This is not to mention that many ERPs have advanced e-commerce functionality, such as forecasting demand and managing the purchase order process. Now, we're ready to share data with other systems.

In the next chapter, we're going to look at using a **point-of-sale** (**POS**) system so that you can sell your products in person.

7
Configuring In-Store POS Solutions

If you have experience in the retail world, you're probably very familiar with the term **POS**. It stands for **Point of Sale**, and it's the system an employee uses to enter an order.

They usually have a simple touch interface where you tap products, add them to the order, and pay for the order. And the POS system usually has the technology to process credit cards. Sometimes, it has more advanced functionality, such as adding a customer to the order so that customer data is synced between your online store and your POS system.

POS systems range in complexity. For a mall kiosk or for selling at a convention, you'll probably want the simplest and most compact POS you can find. A bare-bones POS system will likely have the following two components:

- Credit card reader
- Receipt printer

If you have a department store with dozens of checkout locations, you might want additional functionality, such as the following:

- A built-in cash drawer
- Barcode scanner
- Clock-in/clock-out feature for employees
- Layered permissions where certain functionalities can only be accessed by a manager

Since WooCommerce integrates with so many systems, there are a ton of POS options to choose from. Some of the most popular ones are as follows:

- WooCommerce POS (`https://wcpos.com/`)
- Square
- Lightspeed

There are hundreds more to choose from. Each industry will have its own favorite POS systems based on their unique needs. So, you can do some digging by searching for `{{your industry}} + POS systems`.

In this chapter, we'll cover the following topics:

- Setting up WooCommerce POS
- Setting up Square
- Syncing data in-store and online

By the end of this chapter, you should be able to configure a POS system with your WooCommerce store.

Technical requirements

To use Square later in this chapter, you should sign up for a free account. You can do so here: `https://squareup.com/signup/en-us`.

Setting up WooCommerce POS

WooCommerce POS is a plugin built on top of WordPress. It's similar to WooCommerce, the core plugin, which is free and available on GitHub (`https://github.com/kilbot/WooCommerce-POS`). It also has premium features that you can unlock by buying a license.

We're going to explore the free version that anyone can experiment with. Follow these steps to set up the POS:

1. In your WooCommerce site, go to **Plugins | Add New**.
2. Search for `WooCommerce POS`. The plugin is shown in the following screenshot:

Figure 7.1: WooCommerce POS on WordPress.org

3.   Now, you can install and activate the plugin.

Testing WooCommerce POS

WooCommerce POS uses the same database as your WooCommerce site. This is great in many ways but it also means that if you experiment with WooCommerce POS, it's easy to clutter up your database. So, if you experiment with this POS system, I recommend you do so with a test site.

Once you've installed the plugin, you'll see a new item called **POS** in your WordPress admin menu. From there, you can immediately see what it does by clicking **POS | View POS**. Here's what it looks like on the screen:

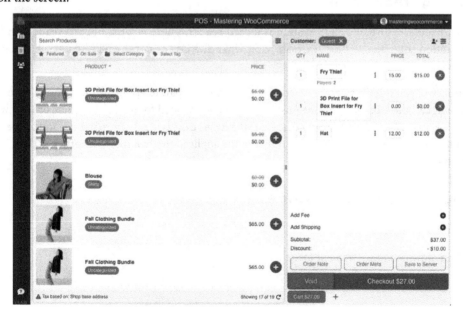

Figure 7.2: Using WooCommerce POS

Since this plugin is built directly on top of WordPress, with just a couple of clicks, we can see the results. We can see a list of products in three buckets:

- **All (default)**
- **Featured**
- **On Sale**

We can also search by a specific category or tag. Once we've narrowed down our list of products, we simply click on the product, and the items are added to our cart on the right. We can add a fee, add shipping, and assign this order to a specific customer.

Then, we can click **Checkout**. This is where the platform is a bit limited. Since it lives in the browser, it doesn't have a direct connection to a credit card reader. By default, you can check out with either of the following:

- Cash
- Card

If you're paying with cash, it's very easy. Enter the cash received and the order will be marked as paid. For conventions or farmers' markets, this is super easy. Credit cards are a little more challenging. Let's look at how you can accept credit cards and when you should choose WooCommerce POS.

Accepting credit cards

The most useful feature of a POS system is a card reader. There is a list of card readers in the documentation (`https://faq.wcpos.com/hardware/compatible-credit-card-readers`).

To summarize, you'll have to enter the order total into a separate application. If you're going to accept credit cards at a brick-and-mortar store, it is recommended to use an old phone or iPad with Payment for Stripe (`https://paymentforstripe.com/card-readers`). You swipe the card or manually type in the credit card number in the application, then mark the order as paid. Let's see how to use Stripe.

Setting up payment for Stripe

Follow these steps:

1. First, download the *Payment for Stripe* app on your phone.

Open the app and enter the order total, which you can see in the following screenshot:

Figure 7.3: Capturing the payment with Payment for Stripe

1. Click **Next** to continue.
2. On the following screen, enter the credit card information or scan a credit card:

Figure 7.4: Entering credit card details in Payment for Stripe

3. When you're done, press **CHARGE** (currently grayed out in the screenshot), and the funds will be transferred.

Now you can go back to WooCommerce POS, select **Card** as a payment method, and click **Process Payment**.

That's it. If you like this process, you should consider selecting WooCommerce POS as your POS system.

Selecting WooCommerce POS

WooCommerce POS is a little clunky when processing cards. It's obviously nicer to have a built-in terminal that knows the total (so you don't have to re-enter the amount) and attaches all of the relevant details of the order to the transaction, which makes exchanges and refunds easier. WooCommerce POS is *incredibly* easy to set up. You don't need any extra accounts, and you can start by manually typing in credit card details immediately, which means you don't have to get any hardware or pay for *anything*.

WooCommerce POS is the only 100% free option. If you like it, you'll probably want to get the credit card reader, even though it has a one-off fee.

If you want a more comprehensive service, then you'll want to look into Square, which we're covering next.

Setting up Square

If you've stepped into a coffee shop in the last decade, you've probably used Square. It's one of the most popular and general-purpose POS systems. It's designed for small businesses and excels at that role. It has competitive pricing and integrations with major platforms, and the product itself is really easy to use.

We'll look into how we can connect our store with Square, set up the integration, and sync data between Square and WooCommerce.

Connecting with Square

To integrate with Square, you need a Square account, and you need to download the free Square extension for WooCommerce (`https://woocommerce.com/products/square/`). Let's start by installing the Square extension for WooCommerce. Once you've installed and activated the plugin on your site, you'll see a notice to configure it, which you can see here:

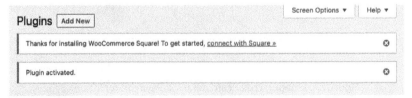

Figure 7.5: Active WooCommerce Square

Follow these steps to configure Square:

1. Click the **Connect with Square** link. You'll be redirected to the **Square for WooCommerce** settings page located at **WooCommerce | Settings | Square**:

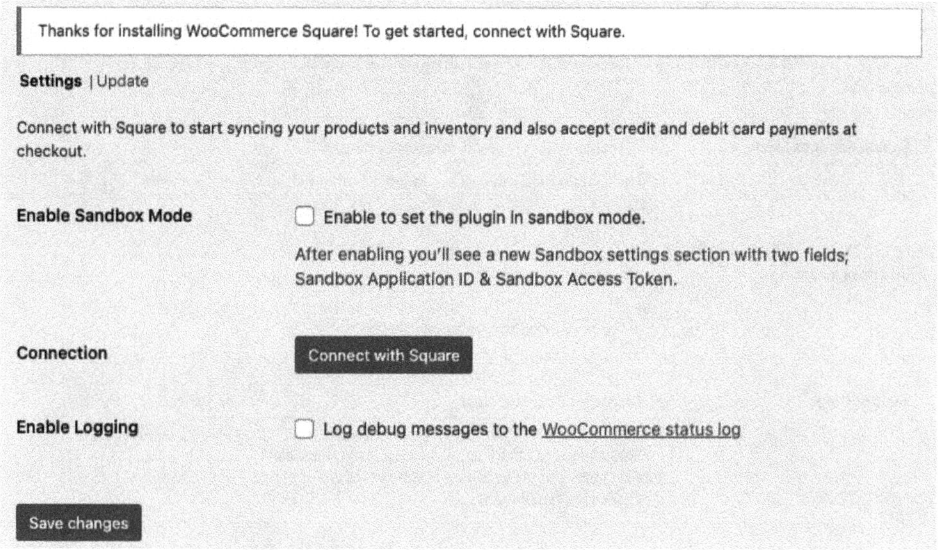

Thanks for installing WooCommerce Square! To get started, connect with Square.

Settings | Update

Connect with Square to start syncing your products and inventory and also accept credit and debit card payments at checkout.

Enable Sandbox Mode   ☐ Enable to set the plugin in sandbox mode.

After enabling you'll see a new Sandbox settings section with two fields; Sandbox Application ID & Sandbox Access Token.

Connection   Connect with Square

Enable Logging   ☐ Log debug messages to the WooCommerce status log

Save changes

Figure 7.6: Connecting WooCommerce to Square

On this page, you'll see a big button to connect with Square.

1. Click on **Connect with Square** and you'll be redirected to your Square account.
2. Login if you haven't already and you should see a page that describes the permissions that WooCommerce needs to interact with Square.

The list of permissions is huge. WooCommerce is basically going to manage every aspect of your Square store. To set up permissions, follow these steps:

1. Click **Allow**.

2.   On the confirmation page, click the **THAT'S MY SITE - REDIRECT ME** button to be redirected back to your WooCommerce store:

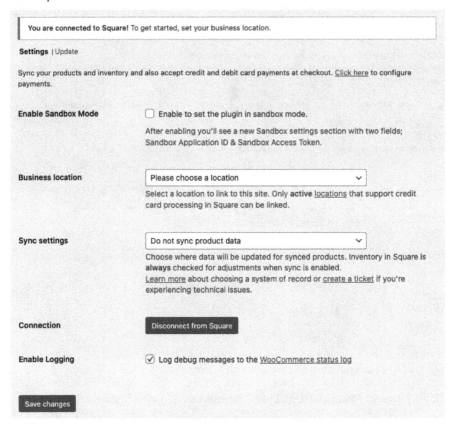

Figure 7.7: Square settings

I had some trouble connecting Square to my WooCommerce site. If you are still seeing **Connect with Square**, wait a few minutes and try again. Sometimes, having to log in, reset passwords, and get two-factor authentication can mess up an integration.

Let's see how to set up Square for WooCommerce in the next section.

Setting up Square for WooCommerce

Now that we've installed and connected Square, we need to adjust a few settings. There are two important aspects that we'll look into:

- **Business location**
- **Sync settings**

Business location is a feature of Square that is designed for multiple physical locations (`https://squareup.com/help/us/en/article/5580-manage-multiple-locations-with-square`). But it can actually be useful for online stores.

If you store your online store inventory in a warehouse, you can create a new location for that warehouse, and it will have a separate inventory from your other stores. Whatever location you select, that is the inventory your WooCommerce store will use.

If everything is stored in one place, then you don't need to create a new location. You can just select the one location that you created when you signed up for Square.

Sync settings (formerly the product system of the record) (`https://woocommerce.com/document/woocommerce-square/sync-settings/`) sounds very technical, but it's actually very simple. Since you have two databases (WooCommerce and Square), WooCommerce needs to know which one is the master.

You can choose from the following three options:

- **Do not sync product data**
- **WooCommerce**
- **Square**

Do not sync product data is the most straightforward. You can have two entirely independent systems. If you don't have a lot of overlap between your online store and your brick-and-mortar stores, this might be a good option for you.

The next option is **WooCommerce**. When WooCommerce is the system of record, your Square product catalog will be overwritten with data from your WooCommerce store for these fields:

- Product name
- Product price
- Inventory count (if **inventory sync** is enabled)
- Product category
- Product image

The last option is **Square**. When Square is the system of record, your WooCommerce data will be overwritten with data from Square for the following fields:

- Product name
- Product price
- Product description
- Inventory count (if **inventory sync** is enabled)

- Product category
- Product image (if a featured image is not set in WooCommerce)

> **Note**
>
> The system of record only needs to be set once. You don't need to set this setting on a per-product basis.

Once the integration is configured, you'll need to know how the data is synced between the systems.

Syncing data

As we mentioned earlier, because we're using two systems, there are two databases: one for WooCommerce and one for Square. It's important to know which system is the master and plays the role of updating the other system.

The option you choose determines how often your data syncs. If Square is the system of record, it updates your WooCommerce store every 60 minutes. If WooCommerce is the system of record, it updates Square immediately, so if a customer places an order, WooCommerce will immediately update the inventory in Square.

Next, we're going to look into how to mark products to sync, when you want to manually sync products, and how the category structures differ.

Marking products to sync

Once you have decided how you want to sync data, you have to mark the products you wish to sync and add a **Stock Keeping Unit (SKU)** to the products (on the **Inventory** tab).

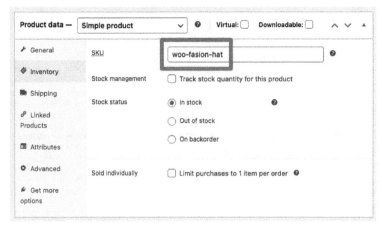

Figure 7.8: Setting an SKU

Once you have an SKU set under the **General** settings for a product, you can select **Sync with Square**.

Figure 7.9: Syncing a product with Square

> **Note**
>
> The SKUs in both systems must match. Otherwise, there will be no data sync.
>
> Furthermore, the WooCommerce website clearly states the following:
>
> "*Unlike WooCommerce, Square does not support multiple attributes. If you would like to sync multiple attributes from WooCommerce to Square, consider combining attributes into one product variation – for example, instead of a "color" attribute and a "size" attribute (e.g., black or blue and small or large), combine the two in your product variations (e.g. black-small, black-large, blue-small, blue-large).*"

Manually syncing products

You can always perform a manual sync. Under **WooCommerce | Settings | Square**, click on **Update** in the submenu, as shown in the following screenshot:

Figure 7.10: Manually syncing products between Square and WooCommerce

Here, you can see all the products marked to sync and the last time they were synced.

There's a handy log that notes what happened with each sync. If a product doesn't sync for some reason, you can see it listed here, as shown in the following figure:

Figure 7.11: Sync history

For the most part, you shouldn't need this, but it's always nice to know there's a log in case something doesn't work correctly.

Flat versus hierarchical categories

WooCommerce and Square are very similar but they do have some differences that are challenging to reconcile. One of those differences is that Square's category structure is flat, like this:

- T-shirts
- Long-sleeved shirts
- Pants
- Shorts
- Capris

However, WooCommerce's category structure has an added dimension. WooCommerce uses a **hierarchical structure**, like this:

- Shirts:
  - T-shirts
  - Long-sleeve shirts
- Pants
- Shorts:
  - Capris

On paper, this looks like a small difference but it creates huge technical challenges. So, the integration is limited for these reasons.

Furthermore, Square only allows one category per product, so if your products have multiple categories and sub-categories in WooCommerce, only the first or parent category will be sent to Square. So, if you really want to get the most out of your integration with Square, you may need to reimagine your category structure so that it works even if it's flat.

Selecting Square

Setting up Square for your WooCommerce site is pretty simple. You can easily sync products and inventory between both systems. You won't have a bunch of canceled orders that you couldn't fulfill due to running out of inventory in your brick-and-mortar store.

The biggest benefit of Square for many WooCommerce site owners is, of course, the POS system. After syncing all of your data, you can use your Square terminal to accept orders at a real physical brick-and-mortar location, or even take it to a weekend-long conference.

The Square for WooCommerce integration is free. Setting up a Square account is also free. With their free plan, they only take commissions on payments so it's risk-free to see whether it works for you. You can see the full details here: `https://squareup.com/us/en/point-of-sale/software/pricing`.

Syncing data in-store and with WooCommerce

In this chapter, we looked at two very different POS systems and learned how they work differently. WooCommerce POS is simple and fast to set up, whereas Square has a lot more flexibility when it comes to which data you want to sync, how often you want to sync, and which database should be the master. As your e-commerce business grows, you'll integrate with dozens of different systems. One of the WooCommerce stores I worked on recently has close to 100 integrations with other systems and they also need to share product data with two other WooCommerce stores. When systems get that complex, it's worth looking into all of the ways you can sync data so you understand which works best for you and your business. We'll look into simple systems such as a single database, having one master database, and manually syncing data.

Single database systems

One of the most robust systems you can use is a single database. This is one of the huge advantages of WordPress. With WordPress, there's only one database for WordPress data, your plugin data, and your theme data. You could have a WooCommerce plugin that offers a post-purchase upgrade. That plugin can display such an offer because it's accessing the order data that was just saved to the database.

From a data accuracy perspective, everything is always up to date. As soon as you have multiple databases, it's possible to have inconsistent data. As an example, let's say you have a POS system that doesn't integrate with WooCommerce. If you signed up for a loyalty program in-store at the POS system, you will be invisible to WooCommerce and loyal customers won't be able to use their rewards on your WooCommerce website.

WordPress starts as a single database system. If you can maintain just that single database, your data will always be available to itself.

Mastering synced databases via an API

If you have two separate websites that talk to each other (such as Square and your website), there's probably going to be two separate databases talking to each other. When two systems communicate, it's often via an API. One system will be able to send out a notification that something happened, such as a new order in Square. WooCommerce could respond to that new order information and do whatever your website is programmed to do, such as creating a new customer record, saving the order data, and decreasing product inventory.

In that case, you'll need to have *one source of truth*. One source of truth is an expression in the programming world to stress how important it is to have a primary database and sync all other data from that one source.

If you have multiple sources of truth, there's very likely to be some data loss at some point when syncing. You can have WooCommerce as your one source of truth or another system. It's your choice, and it really comes down to how much you like working with the different systems.

There are sometimes minor incompatibility issues, such as the different category structures in WooCommerce and Square. However, for the most part, data should flow pretty easily between both systems.

Manually syncing data

You can, of course, export a CSV from one platform and import it into another platform. I've had to do this on some projects while we were building an automated integration and it gets *very* tedious, very fast. And there's always the opportunity for human error.

I don't recommend manually syncing data for an extended period of time. You will eventually forget to sync or incorrectly sync data and have messed up orders, which can be a huge problem for an e-commerce store.

Summary

In this chapter, we learned how to set up two different POS systems: WooCommerce POS, which is built on top of WordPress, and Square, which is an independent system with a ton of settings. We also looked into how you can best sync data between your WordPress site and other systems.

You should now be familiar with the basics of setting up a POS system for your WooCommerce store. This will allow you to sell your products in person at a convention or in a brick-and-mortar store. As you complete orders in one system, they'll be synced to the other system, keeping all of your inventories in sync and reporting accurately while also making new product launches no more difficult than creating a product in WooCommerce.

Depending on the POS system you choose, you might also be able to manage hourly employees who clock in and out of their shifts. There's a lot of power in connecting your WooCommerce store to a POS system. Now that you know how to do this, your business can build an organic physical presence.

In the next chapter, we're going to talk about using fulfillment software to make sending orders much easier.

8

Using Fulfillment Software

In *Chapter 5, Managing Sales Through WP-Admin*, we talked about managing orders, including how to mark orders as shipped and how to print labels through your site. That's the easiest way to fulfill orders and that's how I recommend most store owners start.

Once your store starts growing and you need to ship out dozens of packages a day and you have full-time employees helping you pack and send orders, you probably want to set up software to help you ship packages. This is called fulfillment software. Alternatively, you might hire a **third-party logistics partner** (**3PL**) who will ship your packages for you. Both options have numerous advantages. Here are a few examples:

- You can easily give employees access to fulfillment software so that they can do everything in their job (picking and packing items, printing shipping labels, and marking orders as complete) without giving them access to your whole site.

- Most fulfillment software provides built-in shipping discounts, meaning you'll save money using this software rather than dropping off packages at the local post office.

- Combine orders from multiple online stores and marketplaces in one place to make them easy to fulfill.

- 3PLs will do the work for you. You can contract with a 3PL instead of having to lease a warehouse and hire staff to pack and ship the orders.

Now that the advantages of using fulfillment software are clear, let's dive deeper into this topic. Throughout this chapter, we're going to set up several different ways of fulfilling orders. By the end, you should have an easy-to-use system customized to your WooCommerce store and business to get packages out the door.

The following topics will be covered in this chapter:

- Sending and updating shipping information
- Configuring Shippo
- Configuring Shipstation

Let's start with sending and updating shipping information in WooCommerce.

Sending and updating shipping information

There are hundreds, if not thousands, of fulfillment companies. There are a few very well-known companies, but there are also hundreds of local fulfillment companies.

The bigger companies that have direct integrations with WooCommerce should automatically retrieve and update order data, but if you're working with a smaller company, it's important to know what you have to do to send them data and update data on your end.

We're going to be looking into how you send shipping information to any company and how you can update your shipping data.

Sending shipping data

Fulfillment companies need your data to do their job. They need to know who the order is being sent to, their address, what goes in the order, and how the business wants to ship the order. Without this data, they can't do their job and your packages will just sit in their warehouse.

Some of the biggest fulfillment companies, such as Shippo and Shipstation, have direct integrations with WooCommerce, in which case you can install the direct integration and you should be good to go. But if you don't have a direct integration, what can you do?

We're going to look into two solutions (Shippo and Shipstation) that don't require any code, and anyone can implement them. We'll also provide a high-level overview of how you could build a custom integration.

Sending emails

One of the lowest-tech solutions is to send an email to a fulfillment provider. Some 3PLs allow you to send an email with order details. These companies are equipped with software to parse your email searching for important data. They just need the email to include the following details:

- The customer's shipping address and phone number
- Order number
- SKUs and quantities
- How you want to ship the order (such as priority mail or overnight)

This is a low-tech solution and something you're more likely to see at a local logistics company. They'll keep track of the labor and send you a bill every month. What's great is that you don't have to customize your emails from WooCommerce. They already have all of the essential information.

You can go to **WooCommerce | Settings | Emails | New Order** and then click **Manage**. From here, you can CC as many email addresses as you like. This can be seen in the following screenshot:

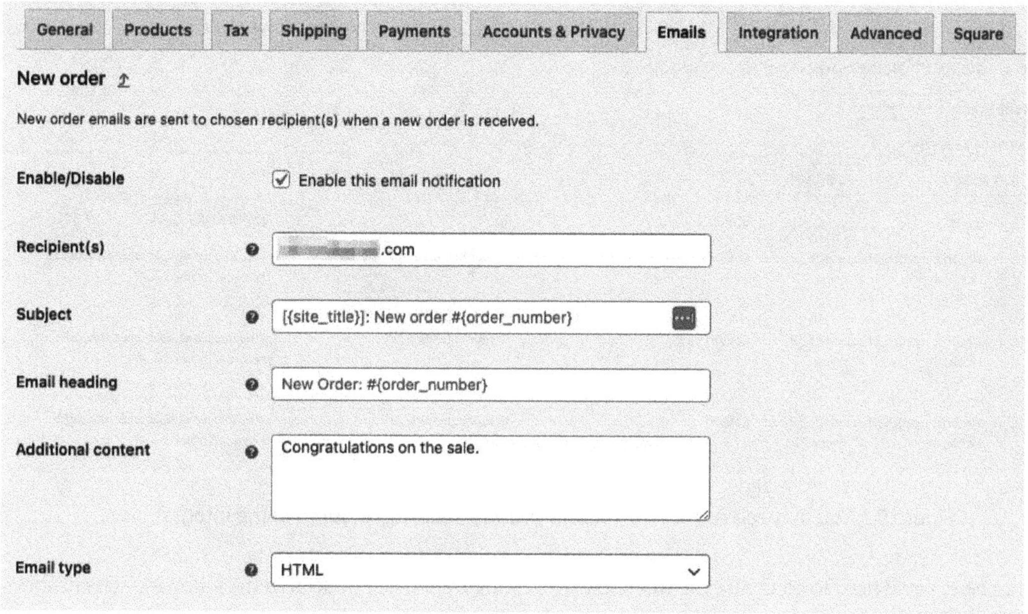

Figure 8.1: WooCommerce email settings

If your fulfillment company needs help with parsing your email, you can change the email type from HTML to text. This will help them dissect the email to pull out the relevant details.

When you're done, click **Save**; all of the emails will be sent to your fulfillment company's admin.

Configuring webhooks

A solution that is a bit more programmatic and reliable is using **webhooks**. If you haven't used webhooks before, they're basically a notification system. So, whenever a specific action happens on your site, a webhook will send some data to a destination. In our case, whenever an order is placed, we want to send the order data to a fulfillment partner.

How is this different from an email? Webhooks send data formatted in a specific way that any programming language can parse automatically. This is in contrast to emails, which are formatted for humans. Email parsing programs are complex and can make mistakes. Webhooks are made for programming languages and are far more reliable.

If you have the choice between sending an email and using a webhook, I would much rather send data via webhook.

You can go to **WooCommerce | Settings | Advanced | Webhooks** to get started. Then, click the **Add webhook** button. This can be seen in the following screenshot:

Figure 8.2: You may see some webhooks if you've already configured some integrations

From here, you'll need to configure the webhook. The configuration can be seen in the following screenshot:

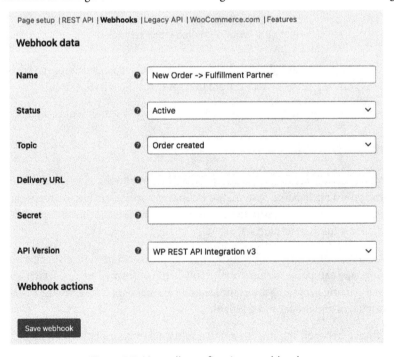

Figure 8.3: Manually configuring a webhook

You'll need to give it a name to help you remember what the webhook does. Set **Status** to **Active**; for **Topic**, you'll want to set it to **Order created**.

You'll need to get the **Delivery URL** and **Secret** values from your logistics partner. Once you've done that, click **Save webhook** – your data will automatically be sent to the partner.

Don't forget to get the **Delivery URL** value from your fulfillment partner. They have to provide an inbox for this data; otherwise, the webhook won't do anything. Once you save the webhook, all new order data will be sent to your fulfillment partner and they can begin fulfilling orders immediately.

There is one downside to webhooks: they only send the data *once*. If your fulfillment company's system is down for maintenance or some other reason, they might miss the notification. It's rare but it can happen, and it's why you might want to look into a custom integration.

Building a custom integration

One of the best things about WooCommerce is that it's open source technology, and you can easily build integrations. Building an integration means you're writing code to interact with WooCommerce's code and you're sending that information to a new destination.

You can build a solution that sends order data to a fulfillment partner's **application programming interface (API)**. This is what the major shipping logistics companies do. You can browse through the code in the Shippo and Shipstation plugins to see how they send data to their APIs.

This project will likely take several days or more likely a week. It is beyond the scope of this book to go into the details of this project. However, there are plenty of resources on building WooCommerce extensions. Here is one such example: `https://saucal.com/blog/woocommerce-custom-plugin/`.

Building an integration is an ambitious project. It doesn't take much code, but you must be comfortable enough with WordPress and WooCommerce to dig into the code to know exactly what you need to write. It is the most robust process. If you plan on building your livelihood on your store, you should – at some point – build a robust integration.

Updating data

We've talked about all the ways you can send data to logistics partners, but we didn't talk about how you can update data on your site. Some store owners don't update their data. Instead, they send order information off to a 3PL and hope they ship everything correctly.

This *can* work, but when customers write in with their questions about their missing order, how do you handle that? We have to rely on a third party who may or may not be great at responding to email.

It's a good idea to track all the data on your site as well.

At a minimum, you'll want to track the following:

- Order number
- Order status
- Tracking number

This data will help you answer any customer service question. Let's talk about two different ways you can get this information from a third party.

Processing a daily email

The lowest-tech solution is that some fulfillment companies will send you a daily email with the orders that have been fulfilled, along with their tracking numbers.

You could manually update this information in WooCommerce. Depending on how many orders are shipped a day, this could be a big-time commitment.

However, the benefits outweigh the time commitment because, when you have this data in one place, you can answer any customer support question and you will have access to better/more useful reports through WooCommerce.

Retrieving order data through a custom integration

An idea to save you a lot of manual processing time and reduce human error is to update your custom integration. If the logistics partner has an API, you can write an integration to check the status of your orders every hour and update them in WooCommerce.

This will take some time to build but will save you countless hours throughout the year. And again, this is one of the main advantages of WooCommerce. Everything is open and customizable; you can build all the tools you need.

With that, we've looked at several low-tech as well as high-tech solutions for sending and receiving data from fulfillment partners. If your fulfillment partner doesn't have a direct integration, you'll have to go with one of these. Pick one and get everything connected so you can make your business more efficient.

Now, let's learn how to integrate and print labels for our orders by using Shippo.

Configuring Shippo

Shippo is a platform that helps you quickly print cost-effective shipping labels. While platforms such as WooCommerce have done a good job of optimizing printing shipping labels one at a time, companies such as Shippo, which specialize in shipping, have a whole bunch of extra features:

- Printing batches of shipping labels
- Automating return labels
- Pricing discounts for shipping labels

Most fulfillment software charges a monthly fee, but Shippo has a freemium model, so you can sign up for free, try it out, and see if it's right for you.

We're going to sign up for Shippo and then show you how to fulfill orders with Shippo.

Signing up for Shippo

Shippo makes it easy to sign up for their service. They have a free account option that only takes a moment to set up. Let's set one up. Follow these steps:

1. Start by creating a free account on Shippo.
2. Once you've created your account, Shippo will ask you what platform you're using, as shown in the following screenshot. Click on **WooCommerce**:

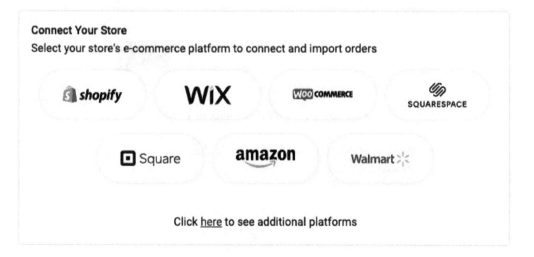

Figure 8.4: Shippo eCommerce platform integrations

3. Enter the URL for your store into the popup from Shippo.

> **Note**
>
> When you're connecting to your store, make sure you use HTTPS or HTTP appropriately. If you try to enter an HTTP URL for a site that requires HTTPS, you might not be able to connect.

4. You'll be redirected to a wizard to help you connect your WooCommerce store to Shippo, as shown in *Figure 8.5*:

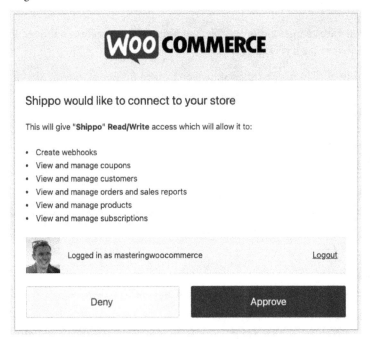

Figure 8.5: Connecting WooCommerce to Shippo

5. Click **Approve** in the confirmation box. Once you do this, your WooCommerce store and Shippo will be connected.

6. Your orders should automatically be imported into Shippo. If they're not, you can always manually import orders by clicking **Sync orders** in Shippo. You can see my imported orders, as well as the **Sync orders** button, in the following screenshot:

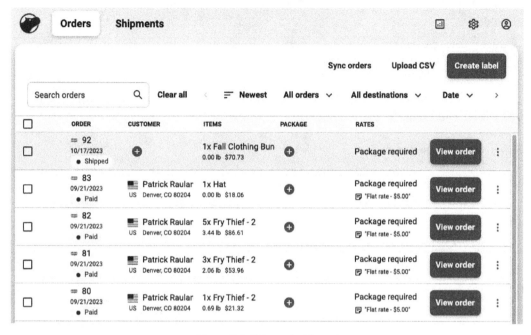

Figure 8.6: Shippo dashboard

> **Note**
>
> If you don't already have orders in your store, you'll want to make some test orders so that you can see orders in Shippo and then continue with this chapter.

Configuring Shippo's setup information

This isn't a book about Shippo, so we're not going to cover every single step. But before you start shipping real orders, you'll want to configure a few more things:

- Default sender address

- Payment information

- Package sizes

You'll need all of these to print shipping labels through Shippo. They have a useful help section that you can use to finish the setup (https://support.goshippo.com/hc/en-us).

Fulfilling orders with Shippo

Once you're fully set up with Shippo, it can make fulfillment *much* faster. As a simple example, if you pack 10 to 20 packages from the previous day, you can greatly speed up your efficiency by printing packing slips.

Let's print packing slips in Shippo. Follow these steps:

1. First, go to your dashboard in Shippo.

2. Click on **Unfulfilled orders** under **All orders**, as shown in the following screenshot. You'll see all of the orders that need to be shipped:

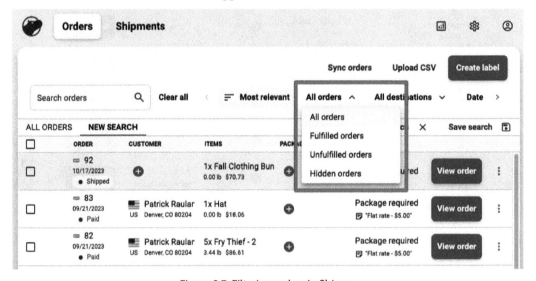

Figure 8.7: Filtering orders in Shippo

3. Select **Unfulfilled orders**; you'll see a new option at the top of the screen under the **...** menu. Click **Download packing slip**, as shown in the following screenshot:

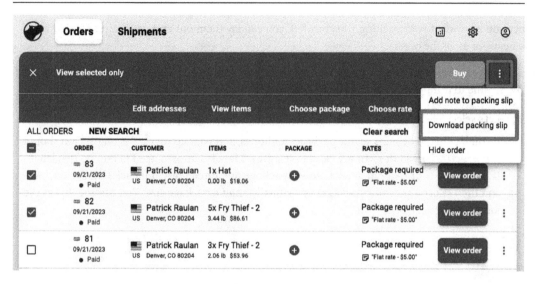

Figure 8.8: Printing packing slips

4. Click **Download Packing Slip**; you'll get a PDF containing all packing slips. It will look like this:

Packing Slip for Order 82

Patrick Rauland   Order ID: 82
                                                       Order Date: 09/21/2023

                              Ship To: Patrick Rauland
                                        123 Main St.
                                        Denver, CO 80204
                                        US

ITEMS	QTY	WEIGHT	PRICE	TOTAL
Fry Thief - 2 FRY-001	5	0.69 lb	$ 15.00	$ 75.00

			Subtotal:	$ 75.00
			Shipping:	$ 5.00
			Total:	$ 86.61

Figure 8.9: A packing slip from Shippo

Once you have all of your packing slips printed, you can lay them out and use them to gather all of the items for all of the orders.

Once you've gathered all of your items and put them into your boxes with the help of your packing slips, it's time to print your shipping labels. Follow these steps:

1. Let's select all shipping orders, just like we did with packing slips.

2. If you have good defaults set for packages and rates, and your credit card is on file, you should see an active **Buy** button. In my case, I see a **Buy 2 labels** button since I'm buying two labels. If the button is grayed out, click **Choose package** or **Choose rate**. I'll click **Choose package** and then select one of our packages or manually enter a package size. You can see this in the following screenshot:

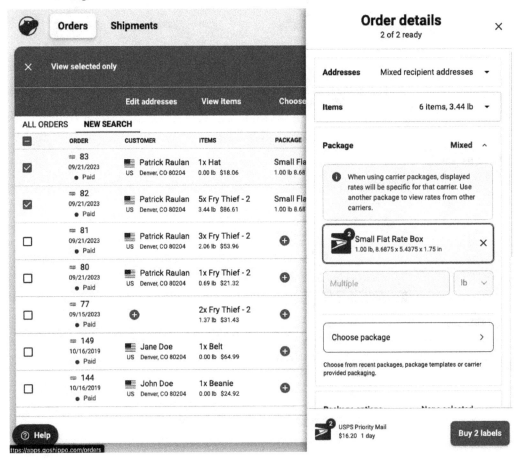

Figure 8.10: Buying shipping labels in Shippo

3.   Either on this screen click **Buy X labels** or go back to the previous screen and click **Buy X**.

Once you've done this, all orders will be updated in Shippo and all shipping information will be sent back to your WooCommerce site.

With that, we've learned how you can connect our store to Shippo, how we can easily view unfulfilled orders, how to print out packing slips to gather goods to be shipped, and how to fulfill those orders with Shippo.

Now that we know how to print labels using Shippo, let's look into one more popular fulfillment company called ShipStation.

Configuring ShipStation

In this section, we're going to configure one of the more robust fulfillment platforms, which is ShipStation. This is a fulfillment company that offers similar functionality to Shippo. Their software is more complex. That means it's more powerful and you can do more with it, but it's a little harder to learn to use. They also have a convenient app that makes it easier to fulfill things on the fly.

If you want to have some of the best fulfillment software in the industry and you're willing to pay for it, you'll want to look into ShipStation.

We're going to set up ShipStation to handle our store's fulfillment. Specifically, we're going to do the following:

- Integrate our WooCommerce store with ShipStation
- Print packing lists
- Print shipping labels
- Use the Shipstation app

The first thing we have to do is integrate our store with ShipStation.

Integrating with ShipStation

ShipStation has a 30-day free trial that you can use to get started. Once you sign up for an account, you'll have to do the following:

- Integrate your selling channel (your store)
- Add your ship from a location
- Select a shipping carrier
- Configure your label layout

We can see some of these options in the following screenshot:

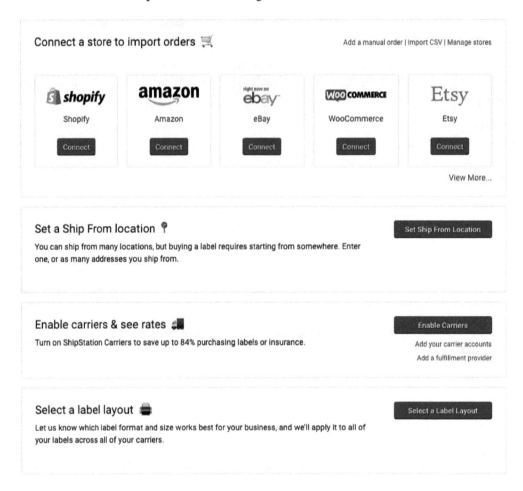

Figure 8.11: ShipStation onboarding wizard

We'll start by integrating our online store, which is sometimes called a **selling channel** in ShipStation. Follow these steps:

1. Click **Connect**, as shown in *Figure 8.12*:

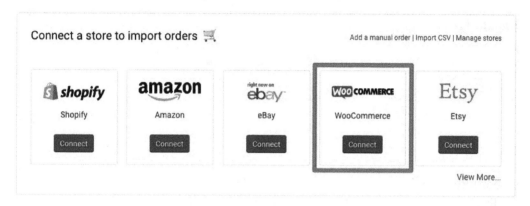

Figure 8.12: Connecting WooCommerce

2. Download the **ShipStation for WooCommerce** plugin from WooCommerce, as shown in the following screenshot:

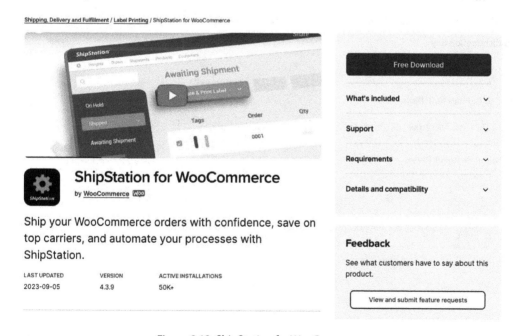

Figure 8.13: ShipStation for WooCommerce

3. Upload the plugin to your WooCommerce site through the plugin uploader.

    Activate the plugin; you'll see the following screen:

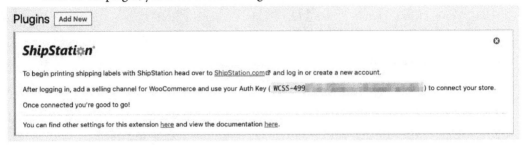

Figure 8.14: ShipStation for WooCommerce activated

4. Copy the **Auth Key** value on that page into ShipStation.

5. Then, add your URL to ShipStation, as shown in the following screenshot:

    5. Copy the **Authentication Key** from WooCommerce and paste it below.
    6. Enter your WordPress site's domain (e.g., http://www.yoursite.com) into the field below.
    7. Within WooCommerce, choose which order statuses to export to ShipStation and map these statuses by placing them in the relevant fields below. Be sure to separate multiple statuses with a comma.
    8. When finished, click Save Changes in WooCommerce and then the Connect button below.

    This plug in requires the use of WooCommerce 2.2+. For a more detailed setup guide, please click here.

    Authentication Key

    ••••••••••••••••••••••••••••••••••••

    URL to Custom XML Page

    https://masteringwoocommerce.mystagingwebsite.com/

    Awaiting Payment Status

    Pending

Figure 8.15: Adding your authentication key and URL to Shipstation

6.   Click **Connect**; you should see a success message, like this:

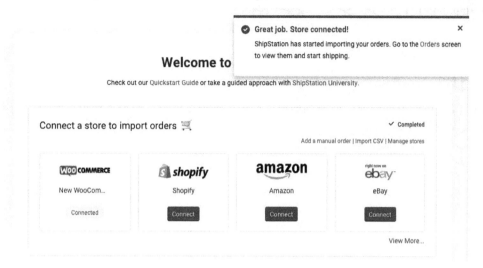

Figure 8.16: Store connected to ShipStation

From here, you can add more sales channels (additional WooCommerce stores, Amazon, Etsy, and so on) or continue to set up ShipStation.

We won't cover every aspect of ShipStation, but you will have to configure a specific shipping carrier, configure your shipping labels, and add a shipping address.

Fulfilling packages with ShipStation

Once you've connected your sales channels with ShipStation, you'll see all of your orders from all of your channels in one place. This can be seen in the following screenshot:

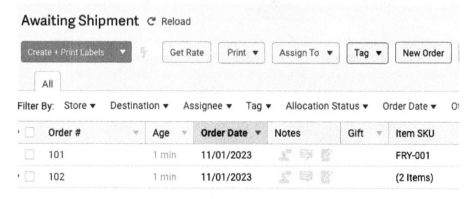

Figure 8.17: Orders in ShipStation

If you click on an order, you will see all the order details. On the right-hand side, you can configure your shipment. You can add the weight, select the package and the service, and then create and print the label. We can see this in the following screenshot:

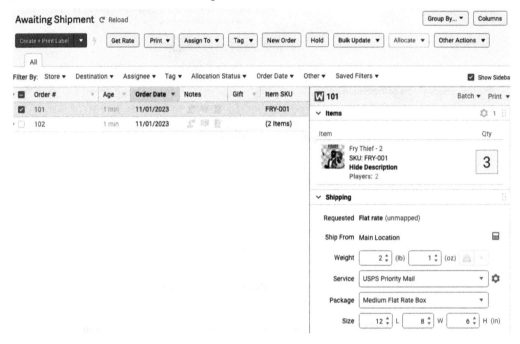

Figure 8.18: Order detail page in ShipStation

Rearranging the order detail panel

ShipStation makes it easy to rearrange the tabs in their interface. I like seeing the items in the order before seeing the shipping details, so I moved the **Items** tab above the **Shipping** tab.

If you select multiple orders, you'll see a different screen where you can validate addresses, which can be super helpful. I've had users give me the wrong address, but this can often be caught with this feature. It prevents orders from going missing, which saves you money, as shown in the following screenshot:

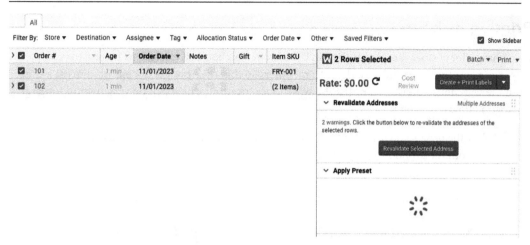

Figure 8.19: Validating addresses in ShipStation

You can also print the packing slips, as shown in *Figure 8.20*:

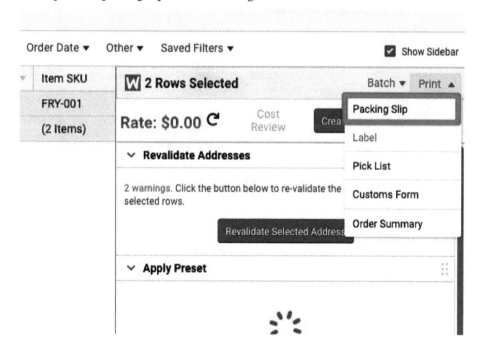

Figure 8.20: Printing a packing slip

Once you've printed the shipping labels, the order should be marked as shipped and all shipment information will be passed back to the original sales channel.

Printing pick lists

If you have a lot of **stock-keeping units** (**SKUs**) in your warehouse, it can also be helpful to print pick lists. You print these out and give them to members so that they can get dozens of items at a time from the warehouse shelves. They list the warehouse location, which can be very helpful for large catalogs. Follow these steps:

1. Select the orders you want to ship. Then, click on **Print** at the top of the page and select **Pick List** from the dropdown, as shown in the following screenshot:

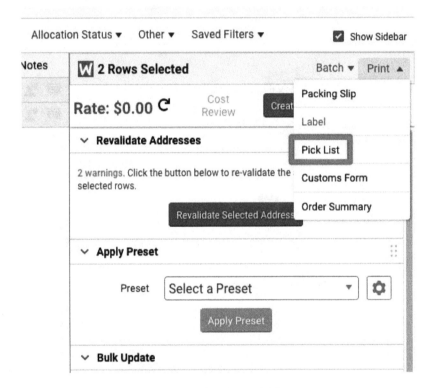

Figure 8.21: Printing a pick list

2. You can then print the pick list. It should look similar to the one shown in *Figure 8.22*:

Product Pick List   Wednesday, November 1, 2023 1:22 PM

	Item #	Description	Warehouse Location	# Required
	FRY-001	Fry Thief - 2		3
	Players: 2			
	woo-fasion-hat	Hat		1
	woo-fasion-jacket	Jacket		1

Total Items Required: 5

Figure 8.22: A pick list helps you pick all of the items that need to be shipped

With this handy slip of paper, we can navigate the warehouse much faster and make fewer trips because there will be fewer errors.

Using the ShipStation app

One of the nice features of ShipStation is that there is an app. If you don't want to print a bunch of paper out every day to ship your packages, you can use the app. You can view orders like so:

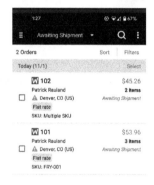

Figure 8.23: Orders in the Shipstation app

You can also select multiple orders and view or print **Packing Slips** and **Pick Lists**:

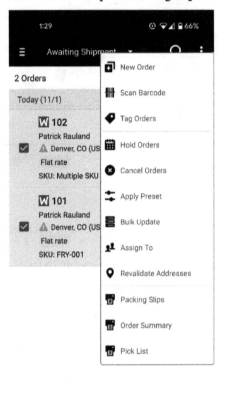

Figure 8.24: You can print pick lists, packing slips, and more from the app

ShipStation generates the same pick list as it does on the website. It's a PDF and you can view that pick list on your device.

ShipStation is an incredibly powerful fulfillment software. If you grow your company to the point where you have a huge warehouse full of inventory and multiple full-time staff, or if you just need additional tools, then ShipStation is a great option.

Summary

Fulfilling orders is a fundamental part of eCommerce. Unless you're only selling digital goods, you'll have to create and optimize your fulfillment process.

You should now be able to send order information to third parties, and you have both high-tech and low-tech solutions. You can integrate Shippo with your online store and fulfill orders through their interface and integrate ShipStation with your online store, create pick lists, fulfill orders, and use their handy mobile app.

With the tools and techniques mentioned in this chapter, handling fulfillment for your store should be much easier.

If you're a brand new store and you're looking at your first fulfillment software, I recommend Shippo as it's a little easier to use. If you are a larger online store that has numerous employees who work in the warehouse, I recommend ShipStation. Their tool is more comprehensive and gives you more control over the fulfillment process.

In the next chapter, we're going to look into speeding up your store.

9
Speeding Up Your Store

When you're running any sort of website, speed is important. But this is even more important with e-commerce where users are browsing pages, adding items to their cart, filling out checkout fields, and, most importantly, paying you.

To illustrate just how important this is, take a look at this statistic: 63% of shoppers bounce when page load times exceed 4 seconds (source: `https://www.ecommercespeedhub.com/site-speed-standard-2022`).

That's over half of your audience! And if you're not careful, your site can take much longer than 4 seconds to load.

Another statistic: a site that loads in 1 second has a conversion rate 3x higher than a site that loads in 5 seconds (source: `https://packt.link/bPo2w`).

Here's a little math to illustrate this point: let's say 10,000 people visit your website to buy a $50 product:

- A 1-second page load time at a 3.05% conversion rate results in $15,250
- A 2-second page load time at a 1.68% conversion rate results in $8,400
- A 3-second page load time at a 1.12% conversion rate results in $5,600
- A 4-second page load time at a 0.67% conversion rate results in $3,350

In the span of ~4 seconds, potential sales have dropped by $11,190!

For these reasons, it's incredibly important to have a website that loads quickly!

In this chapter, we're going to make sure your site loads quickly.

The following topics will be covered in this chapter:

- Monitoring speed and performance
- Minifying CSS and JavaScript resources
- Optimizing images
- Caching and e-commerce
- Optimizing content above the fold

By the end of this chapter, you should know how to monitor the speed of your site and have a variety of techniques to help you speed it up.

Technical requirements

To use GTmetrix later in this chapter, you should sign up for a free account. You can do so here: `https://gtmetrix.com`.

Optimizing a website for performance can be a little technical so some website hosting companies will do some of the work. They might do the following:

- Automatically install plugins
- Use server-level technology to cache information
- Offer their own **content delivery networks (CDNs)** to make it faster to serve static assets such as JavaScript, CSS, and images

Because a website host could entirely change how you optimize your site you should wait to optimize your site until you've chosen a host and launched your site.

Furthermore, some website hosts will prevent some plugins that conflict with their own technology. As an example, my host, Pressable, prevents me from installing **WP Fastest Cache** because it already has technology that automatically caches my website.

You can typically find a list of disallowed plugins on your host's website. Here's the page for Pressable: `https://pressable.com/knowledgebase/disallowed-plugins/`.

Monitoring speed and performance

Before we can start improving the performance of our site, we need to understand how fast our site is loading and where there might be opportunities for improvement. Without these tools, you're just guessing what will speed up your store, and you'll waste a lot of your time. You should also use these tools to track each change you make to ensure it actually speeds up your store.

In this book, we're going to use **GTmetrix**. GTmetrix is a free tool that shows you exactly how fast your page loads and it also shows where any speed issues might come from. We're going to create a baseline, test changes, look into waterfall data, and finally, set up periodic reports.

Since the first version of this book was published, PageSpeed Insights (`https://pagespeed.web.dev/`) has released a minimalistic yet powerful, and free, tool to monitor your page speed. You're welcome to use this tool instead of GTmetrix, although later in this chapter we're going to set up weekly alerts, which you can't do with PageSpeed Insights.

Finding a starting point with GTmetrix

Before we start changing our site, let's measure our starting point to see how much work we have ahead of us.

Perform the following steps:

1. Register for a free account on GTmetrix. This will make sure your tests are prioritized over anonymous users, and, more importantly, you'll be able to set up automated tests and compare your speed against previous time periods.

2. Copy the URL for your site.

3. Paste it into the URL field on the GTmetrix home page, as shown in *Figure 9.10*:

Figure 9.1: Add your store URL to GTmetrix

4. Click the **Analyze** button, and it will start analyzing the URL, as seen in the following screenshot:

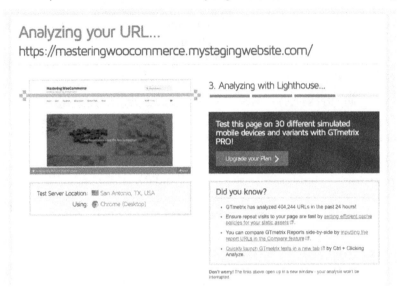

Figure 9.2: It only takes a minute to analyze your store

5. After a minute or so, you should get a report.

There's lots of useful information on the report page. But some of the most useful data is **Web Vitals** and **Speed Visualization**, which can be seen in the following screenshot:

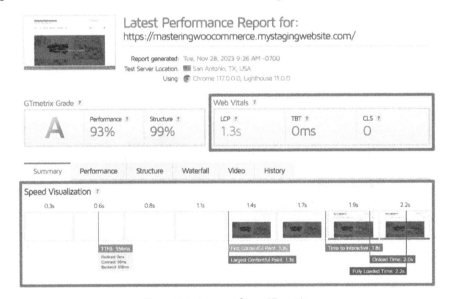

Figure 9.3: A report from GTmetrix

I really like these reports. They provide a high-level overview of the speed of your site in a few metrics while also displaying what your website looks like while it's loading.

GTmetrix Grade

It's incredibly challenging to get a perfect GTmetrix Grade, especially with an e-commerce site. Don't worry about trying to solve every issue. Focus on the biggest issues, and make incremental progress.

Now that we know our baseline for **Largest Contentful Paint (LCP)** is *1.3 seconds*, we can test how new functionality affects our speed.

Web Vitals

GTmetrix and other speed testing tools use a set of metrics called **Web Vitals**. Web Vitals (`https://web.dev/articles/vitals`) were originally created by Google for what they consider "*essential to delivering a great user experience on the web.*"

If you want your website to appear fast to Google and other search engines, you will want to make sure that you have good **Core Web Vitals** scores.

Luckily for us, the Web Vitals team has provided ranges of scores – **Good**, **Needs Improvement**, and **Poor**. Here is a screenshot:

Figure 9.4: Timing from the Web Vitals team

There are currently three metrics:

- **Largest Contentful Paint (LCP)**
- **First Input Delay (FID)**
- **Cumulative Layout Shift (CLS)**

Largest Contentful Paint *(LCP) measures loading performance*

LCP assesses how quickly the largest and most prominent element on a web page becomes visible to the user during the loading process. The general guideline is that the main content should appear within the first 2.5 seconds of the page loading.

First Input Delay *(FID) measures interactivity*

FID is designed to assess how responsive a web page is to user interactions. Specifically, it measures the delay between a user's first interaction with the page, such as clicking a button or tapping a link, and the time the page takes to respond. Aiming for an FID of 100 milliseconds or less is crucial for providing a seamless and interactive user experience.

> TBT versus FID
>
> GTmetrix shows **Total Blocking Time (TBT)** in place of FID because GTmetrix is a simulated test and FID is only available with **Chrome User Experience Reports (CrUX)**. In most cases, TBT is equivalent to FID so treat them the same.

Cumulative Layout Shift *(CLS) measures visual stability*

CLS measures the extent to which page elements shift unexpectedly during the loading experience. For a superior user experience, it is advised to keep the CLS at 0.1 or below.

Testing changes

If we want to add some functionality to our WooCommerce store, we can test it against our baseline (as determined in the *Finding a starting point with GTmetrix* section).

So far in this book, we've built a functional but pretty basic home page in WooCommerce. I have a big banner up top, I highlight my primary product, and I have three accessory products at the bottom with a link to see the whole store. Here is a screenshot of my home page:

Fry Thief

$15.00 – $17.00

Fry Thief is a light-hearted asymmetrical microgame for 2 players. It's a game about poor life choices. Like that time you tried to be healthy and ordered a salad while your friend ordered fries. And how you're going to have "just a few" to make up for your poor life choice.

Players   Choose an option ⌄

Figure 9.5: The home page for Laid Back Games before we make any changes

To add a little more life to my site, I want to add elements to my home page such as additional images and videos. But will that slow down my site? It's hard to know without testing, so let's test it.

I updated my home page with a large testimonial banner, testimonials (in image form), and a few different video reviews, as you can see in *Figure 9.6*:

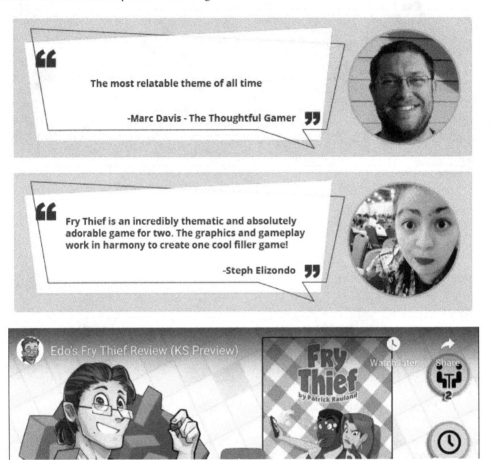

Figure 9.6: Added testimonials and videos on the home page

I honestly don't know the effect this will have, which is why I want to test it. I went into GTmetrix and tested my new home page. The results can be seen in the following screenshot:

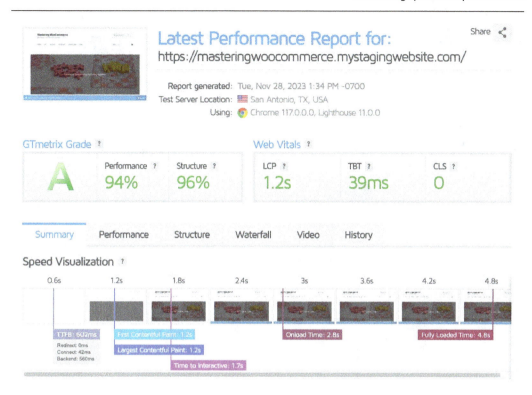

Figure 9.7: A speed test on the improved home page

The total time for the site to load decreased from *1.3* to *1.2* seconds! I did run a few more tests and it's not uncommon to land within *.1* or *.2* seconds, so this is essentially the same speed as it was before. That's a pretty surprising result.

Now I can make an educated decision as to whether the change is worth it. If these changes add 0.5 seconds to my site's speed, will losing a fraction of my audience due to slightly longer load times be offset by the hopefully greater number of people that will convert with a nicer, more helpful page?

In our case, that's a definite yes! We added additional social proof for the potential customers, and it made virtually no difference to our Core Web Vitals.

Now that we've learned how to measure speed, let's discover how to deduce what is causing the speed issue.

Digging into Waterfall data

If you're not happy with the page speed and want to see where the problem is, you can dig into a number of areas in tools such as GTmetrix and PageSpeed Insights. One tool in GTmetrix is the **Waterfall**, and it shows you exactly how each resource is loaded.

Just beneath where you see your performance scores and page details with the fully loaded time and total page size, you'll see a list of tabs.

Click **Waterfall**, and you'll see how each resource is loaded, as follows:

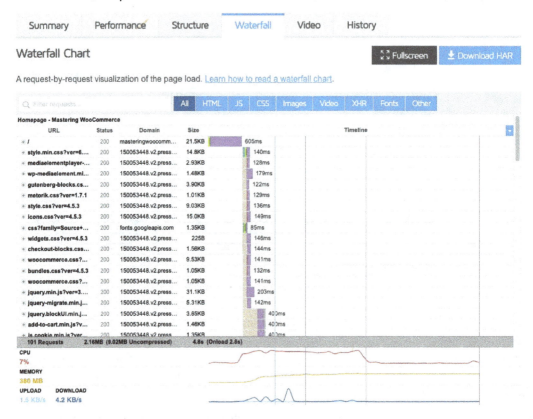

Figure 9.8: A waterfall chart in GTmetrix

Longer bars mean a resource takes longer to load. Keep an eye out for items that take 5-10x longer than their peers.

Slow server response time

You might notice that, before we can load any assets (images, CSS, JS, and so on), we have to load the page itself. In *Figure 9.8*, that's represented by / since it's the root of the site.

The page tells the browser which additional files to download. If this part is slow (taking over 1 or 2 seconds), it's very likely due to a slow or improperly configured server. Contact your host to see how they can help you load the page faster.

In our case, we don't see too many outliers. Everything seems to take a reasonable amount of time to load. The most likely issue you'll see is large images taking a while to load. We'll discuss optimizing images later in this chapter.

Setting up periodic testing

I generally recommend testing your page speed when adding new functionality to your site. That said, it's always nice to have records of your page speed over time. For that reason alone, it's worth setting up automatic periodic testing. Without any prompting from you, GTmetrix will test your sites and compare them against previous records.

Let's configure periodic testing in GTmetrix. Perform the following steps:

1. Click on **Monitor** at the top of the test results page, as shown in the following screenshot:

Figure 9.9: Monitor your website in GTmetrix

GTmetrix free plan limitations

GTmetrix's free plan does have some limitations. You may have to change the testing location. I prefer to use San Antonio to record the best possible speed but I had to change my testing location from San Antonio. If you need specific locations, consider a paid plan.

2.   Choose how often you want GTmetrix to test your site. I recommend either weekly or monthly:

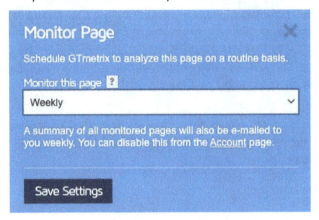

Figure 9.10: Monitor your site daily, weekly, or monthly

3.   Click the **History** tab, and you'll be able to see how the speed of your site changes over time:

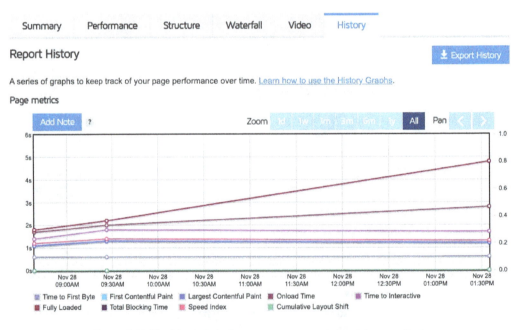

Figure 9.11: The history tab shows your page speed metrics over time

This won't be super useful now since you have only just built the site. But as you gradually add content, update settings, and add new functionality, you'll likely see changes in your page speed, and this tool can help you track how it affects the end user.

> **Monitoring changes within your WordPress site**
>
> If you're on your WordPress site on a regular basis, instead of setting up weekly emails, you could directly pull PageSpeed Insights data into your dashboard using Site Kit (`https://wordpress.org/plugins/google-site-kit/`). This official plugin from Google pulls data from Analytics, Search Console, AdSense, and Speed all into your dashboard, which is powerful.

Now that we know how to track and monitor our site speed, we'll investigate minimizing some resources so our site loads faster.

Minifying CSS and JavaScript resources

One of the easiest ways to get a faster site is to **minify** your files. You can think of minification as similar to reducing the margins on a Word document. You can literally get more words on the page so it takes fewer pages to print. We do that with files such as CSS, JavaScript, and HTML.

There are a few tools that can do that for us:

- WP Rocket (`https://wp-rocket.me`) (paid)
- W3 Total Cache (`https://wordpress.org/plugins/w3-total-cache/`) (free)
- Autoptimize (`https://wordpress.org/plugins/autoptimize/`) (free)

And there are even more than this. There are loads of options to help speed up your site. We're going to use Autoptimize because it's straightforward to set up and there's lots of room to customize advanced settings once you know what you're doing.

Setting up Autoptimize

Let's set up Autoptimize. Perform the following steps:

1. Log in to your site and navigate to **Add Plugins**.

2. Search for **Autoptimize**. Click on **Install Now**:

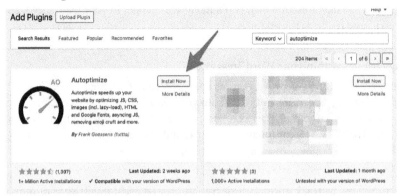

Figure 9.12: Install Autoptimize from WordPress.org

3. Install and activate the plugin.

4. Click on **Settings | Autoptimize** in the **Admin** menu, and you'll see all of the settings.

5. Confirm the checkboxes for optimizing HTML, CSS, and JavaScript are checked as shown in the following screenshot:

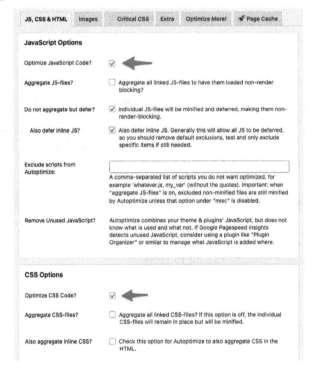

Figure 9.13: Optimize JavaScript, CSS, and HTML files (the HTML checkbox is out of frame)

6.   Click **Save Changes and empty cache**.

Always test optimization plugins

With any sort of optimization plugin, you should test your site after you make any changes. Developers write code in all sorts of different ways, and sometimes a specific plugin or theme can't be optimized automatically; it might break the frontend of your site.

If you do have problems, you can explore the advanced settings or you can only minimize certain files. For example, you could only minimize HTML and CSS without minimizing JavaScript. I tested my site, and all of my assets loaded correctly. So I ran another speed test on GTmetrix. The details can be seen in the following screenshot:

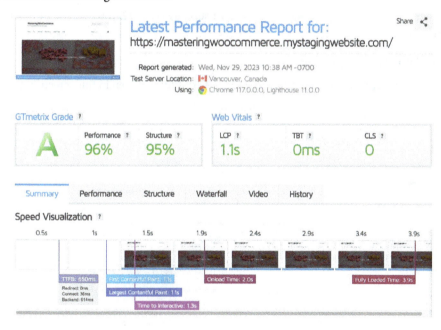

Figure 9.14: Autoptimize improved our performance

Our website was already fast but after reviewing the numbers, Autoptimize was *still* able to improve its performance:

- The LCP went from 1.2 to 1.1 seconds.

- **Time to Interactive** dropped from 1.7 to 1.3 seconds. Users will be able to interact with our website very quickly.

- **Fully Loaded Time** went from 4.8 to 3.9 seconds.

Keep in mind that we're using the exact same HTML, CSS, and JavaScript. They're just served in a format that's more efficient.

Concatenate files if necessary

Technology is always improving and the consensus prior to 2020 was to concatenate files together. Concatenation is where you combine multiple similar files together. So, instead of the browser loading 12 CSS files, there's one concatenated, bigger CSS file to download.

Prior to HTTP/2, this would save time and make your site load faster. Lucky for us, HTTP/2 added a feature called multiplexing, letting browsers download multiple smaller files just as fast as one large file. So, we no longer need to concatenate files together.

However, if you are on an older web server that doesn't support HTTP/2, you may want to know how to concatenate files:

1.   Go to **Settings | Autoptimize**.

2.   Check both **Aggregate JS-files?** and **Aggregate CSS-files?**, as seen in this screenshot:

Figure 9.15: Concatenate files if your server doesn't support HTTP/2

3. Click on **Save Changes and Empty Cache**.

Your files will now be concatenated and served as one large file. Since we're using a modern hosting company with HTTP/2 servers, we don't need to do this.

Now that we've minimized our HTML, CSS, and JavaScript files, it's time to optimize some of the bigger files our site loads: images.

Optimizing images

In the e-commerce world, it's common to hear that you need high-quality photography to highlight your products. And that's definitely true – you do need good imagery. However, you shouldn't simply upload a 2 MB photo to your website. You need to optimize it so it's as small as can be while still being high quality.

This reduces the total size of a page when someone views your product page. This will make your product page load much faster.

We're going to look into two tools to help us do that: **Jetpack** and **Imagify**. First up, we're going to optimize images with Jetpack.

Optimizing images with Jetpack

One of the more well-known tools, and also one that happens to be free, is Jetpack. We already installed Jetpack earlier in this book so it will be pretty easy for us to configure it. Perform the following steps:

1. In WordPress admin, click on **Jetpack | Settings** in the main menu.
2. Scroll down that page, and you should see a list of Jetpack features. Many of which are disabled by default. Scroll down to **Performance & speed**. Check **Enable site accelerator** as shown in the following screenshot:

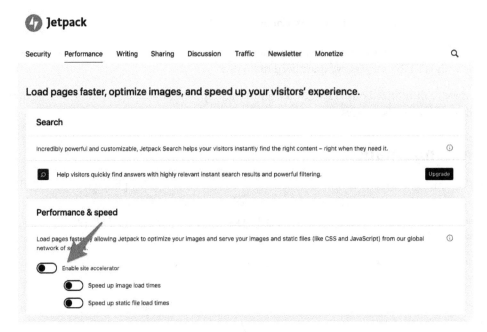

Figure 9.16: Enable site accelerator and the indented checkboxes should be automatically checked

Site accelerator will do the following two things for images:

- **Speed up image load times** will compress images. So a 25 MB photo might only be 400 KB instead.
- It also stores images on `WordPress.com` servers around the world and automatically sends images to visitors from the closest server.

Furthermore, **Speed up static file load times**, will serve common JavaScript files used by WordPress and WooCommerce. So if a visitor sees another site using Jetpack and downloads a `Jquery Javascript` file, they already have it saved in their cache and won't need to download it when they visit your site.

Both of these help speed up your site, and all we did was toggle one setting.

Enabling site accelerator within Jetpack

If you want to test this, then perform the following steps:

1. Go to the frontend of your site and load a product page. It will look as shown in the following screenshot:

Figure 9.17: Our single product page

2. View the source of the page. In Google Chrome, you can do this via **View | Developer | View Source**.

3. You'll see the code, which includes references to WordPress (i0.wp.com). The code is as follows:

```
<div data-thumb="https://i0.wp.com/masteringwoocommerce.
mystagingwebsite.com/wp-content/uploads/2023/07/fry-thief-box-
cover.jpg?fit=65%2C100&amp;ssl=1" data-thumb-alt="Fry Thief
Box Cover" class="woocommerce-product-gallery__image flex-
active-slide" style="width: 323.812px; margin-right: 0px;
float: left; display: block; position: relative; overflow:
hidden;"><a href="https://i0.wp.com/masteringwoocommerce.
mystagingwebsite.com/wp-content/uploads/2023/07/fry-thief-
box-cover.jpg?fit=1158%2C1781&amp;ssl=1"><img width="416"
height="640" src="https://i0.wp.com/masteringwoocommerce.
mystagingwebsite.com/wp-content/uploads/2023/07/fry-thief-
box-cover.jpg?fit=416%2C640&amp;ssl=1" class="wp-post-image"
alt="Fry Thief Box Cover" title="Fry Thief Box Cover" data-
caption="" data-src="https://i0.wp.com/masteringwoocommerce.
mystagingwebsite.com/wp-content/uploads/2023/07/
```

```
fry-thief-box-cover.jpg?fit=1158%2C1781&amp;ssl=1" data-
large_image="https://i0.wp.com/masteringwoocommerce.
mystagingwebsite.com/wp-content/uploads/2023/07/
fry-thief-box-cover.jpg?fit=1158%2C1781&amp;ssl=1" data-
large_image_width="1158" data-large_image_height="1781"
decoding="async" fetchpriority="high" srcset="https://i0.wp.
com/masteringwoocommerce.mystagingwebsite.com/wp-content/
uploads/2023/07/fry-thief-box-cover.jpg?w=1158&amp;ssl=1
1158w, https://i0.wp.com/masteringwoocommerce.
mystagingwebsite.com/wp-content/uploads/2023/07/fry-thief-
box-cover.jpg?resize=416%2C640&amp;ssl=1 416w, https://
i0.wp.com/masteringwoocommerce.mystagingwebsite.
com/wp-content/uploads/2023/07/fry-thief-box-cover.
jpg?resize=195%2C300&amp;ssl=1 195w, https://i0.wp.
com/masteringwoocommerce.mystagingwebsite.com/
wp-content/uploads/2023/07/fry-thief-box-cover.
jpg?resize=666%2C1024&amp;ssl=1 666w, https://
i0.wp.com/masteringwoocommerce.mystagingwebsite.
com/wp-content/uploads/2023/07/fry-thief-box-cover.
jpg?resize=768%2C1181&amp;ssl=1 768w, https://
i0.wp.com/masteringwoocommerce.mystagingwebsite.
com/wp-content/uploads/2023/07/fry-thief-box-cover.
jpg?resize=999%2C1536&amp;ssl=1 999w" sizes="(max-
width: 416px) 100vw, 416px" draggable="false"></a><img
role="presentation" alt="Fry Thief Box Cover" src="https://
i0.wp.com/masteringwoocommerce.mystagingwebsite.
com/wp-content/uploads/2023/07/fry-thief-box-cover.
jpg?fit=1158%2C1781&amp;ssl=1" class="zoomImg" style="position:
absolute; top: -1195.34px; left: -649.191px; opacity: 0;
width: 1158px; height: 1781px; border: none; max-width: none;
max-height: none;"></div>
```

Our images are now being loaded through WordPress.com, and they automatically serve minimized images that speed up our site.

Need to read your source code?

If you're having trouble viewing the source code, you can temporarily turn off Autoptimize. Earlier in this chapter, we minimized our HTML and that can make it hard to read.

Optimizing images with Imagify

Another tool you could use is Imagify (https://imagify.io). This is similar to Jetpack, which compresses and loads your images on its servers to speed up your site. But there are three notable differences:

- It has a free plan and paid plans. You can upload up to 20 MB of photos a month for free, and it'll do everything you need.

- It has more granular control over compression. You can choose exactly how much you want to compress your images.

- It has a **Back** button. You can revert changes at any time and recompress your images.

Let's set this up on our site. Perform the following steps:

1. In WordPress admin, go to **Plugins** and click **Add New**.
2. Search for **Imagify** and click on **Install Now**:

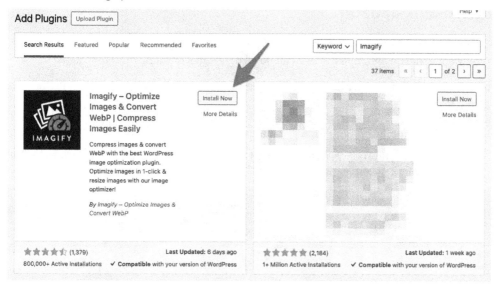

Figure 9.18: Imagify on WordPress.org

3. Activate the plugin.
4. Now we have to connect to Imagify. Start by creating an account if you haven't done so already:

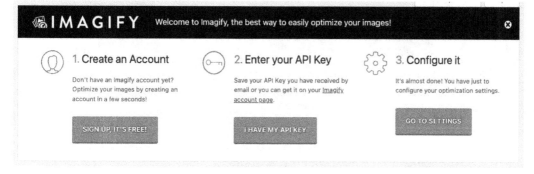

Figure 9.19: Create an Imagify account

5. Once you've created your account, you should get an email with your API key, and you can also find it through Imagify's site. Click on **I HAVE MY API KEY**, and enter it.

And with that, you're done. Images will automatically be optimized, and you'll make your site quite a bit faster. There's also a handy bulk updater we can use.

Using the bulk updater

Optimizing images one by one is great. But sometimes you need to optimize existing content. That's when you want to have a bulk updater that can do all of this in a few quick and easy steps.

One notable feature that Imagify has is its bulk updater. Let's take a quick tour of that. Perform the following steps:

1. In WordPress admin, click on **Media** and then **Library**.

2. You'll see all of your products in a grid. To get the most out of Imagify, click **Switch to the List View** as shown in the following screenshot.

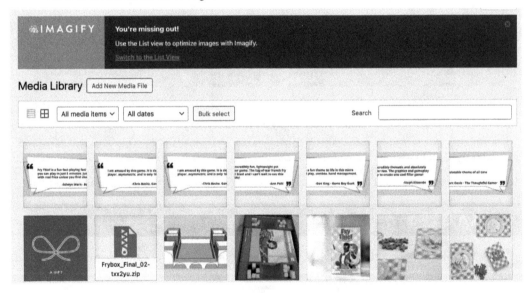

Figure 9.20: Media Library grid view

3. From here, we can optimize single images by clicking **Optimize**. Or we can select multiple images and select **Optimize** under **Bulk actions**:

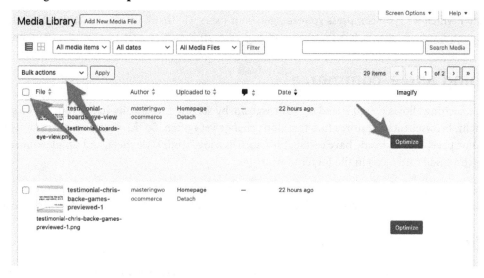

Figure 9.21: Optimize images one at a time or use the bulk editor

4. When you're done, you'll see how much you saved by optimizing each image. And it's quite a bit! This can be seen in the following screenshot:

Figure 9.22: Bulk optimization results – reducing file size by 50-70% is excellent

You can do something similar and get into the details through the **Imagify** menu in the admin bar of your site.

No matter which way you compress your images, your users will thank you for a fast site. Now that we've compressed our images, let's look into caching pages and assets to speed up our site even more.

Caching and e-commerce

Browser caching allows you to speed up your website by storing files locally in the user's browser. Essentially, browser caching stores files that don't change very often. So the next time a visitor visits your home page, they'll already have certain files, such as a logo, your style sheet, and simple elements such as the credit card icons in the footer of your site.

This doesn't help the very first page someone visits, but it will help with each future page they visit on the site. We're going to configure a caching plugin and discuss page caching and when you'd want to use it.

Configuring caching plugins

Let's take advantage of some browser caching. To configure browser caching plugins, perform the following steps:

1. In your admin, under **Add Plugins**, search for **WP Fastest Cache** and click on **Install Now**:

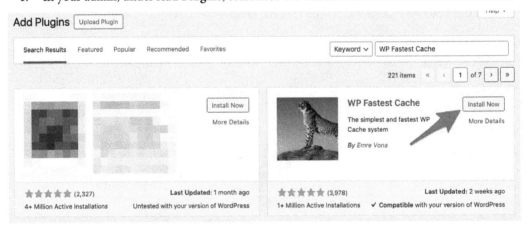

Figure 9.23: WP Fastest Cache on WordPress.org

2. Install and activate the plugin.

3. In your main menu, click on **WP Fastest Cache**.

4. Check the checkbox for **Browser Caching** as shown in the following screenshot:

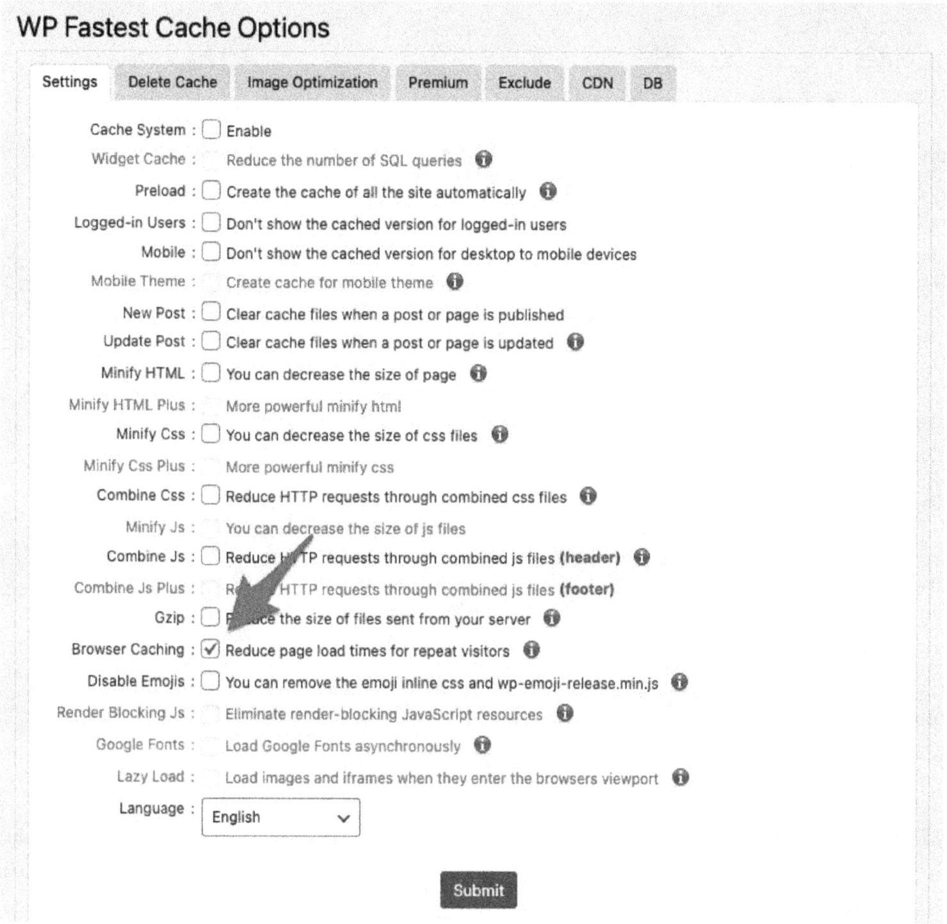

Figure 9.24: WP Fastest Cache settings

5. Click **Submit** to save your changes.

With caching enabled on our site, returning visitors will load the site quicker and it also preserves our web server requests for brand-new visitors. This is a very easy speed win for WooCommerce sites.

Configuring caching via HTACCESS

You can also do this through your `.htaccess` file if you host your site on an Apache server. You'll need to locate your `.htaccess` file on your server.

Note that with most FTP programs, the `.htaccess` file is hidden so you'll need to enable the option to see hidden files.

Then, open that file and add the following code to it:

```
## EXPIRES CACHING ##
<IfModule mod_expires.c>
ExpiresActive On
ExpiresByType image/jpg "access plus 1 month"
ExpiresByType image/jpeg "access plus 1 month"
ExpiresByType image/gif "access plus 1 month"
ExpiresByType image/png "access plus 1 month"
ExpiresByType text/css "access plus 1 month"
ExpiresByType application/pdf "access plus 1 month"
ExpiresByType text/x-javascript "access plus 1 month"
ExpiresByType application/x-shockwave-flash "access plus 1 month"
ExpiresByType image/x-icon "access plus 1 year"
ExpiresDefault "access plus 2 days" </IfModule> ##
EXPIRES CACHING ##
```

If you have the technical expertise, it's a little more efficient to enable this type of caching through `.htaccess` instead of through a plugin, but of course, a plugin is much easier for anyone to install and configure.

Page caching

One thing you should be aware of is the difference between page caching and browser caching. Page caching is where the WordPress site assembles a page (header, main content, images, sidebar, footer, and so on) and saves that page. The next time someone visits that page, the server provides the cached page.

This works great for static sites, such as news sites. But with e-commerce, there are pages you can't cache, such as the cart and checkout. And there are elements within other pages you don't want to cache, such as the cart icon (which usually shows how many items are in your cart) and related products.

Page caching has some challenges with e-commerce. You can attempt to cache some pages, but it's often more trouble than it's worth.

WooCommerce has a list of compatible caching plugins (`https://developer.woocommerce.com/docs/how-to-configure-caching-plugins-for-woocommerce/`) and any modifications you need to make.

Now that we know how to cache our content, we can optimize the content above the fold.

Optimizing content above the fold

On a typical product page, there are a half dozen images. And many of them are "below the fold." The fold is a term from the newspaper industry where there was a literal fold. You wanted your best headlines and photos to be above the fold so everyone would see them.

In the following screenshot, you can see where my fold is on my laptop and what is considered above and below the fold:

Figure 9.25: Above the fold on my laptop

In the web world, we still use the fold terminology but we talk about it in different ways. And when it comes to optimization, we can "lazy load" our images. Lazy loading means we only download the image once the viewer starts scrolling down.

Prior to 2020, you had to use a plugin to lazy load images. However, as is always happening in the technology space, browsers have developed a standard to lazy load images (`https://web.dev/articles/browser-level-image-lazy-loading`) and WordPress supports this standard.

That means your images should automatically lazy load without you having to do anything.

Summary

Optimizing a site can be a full-time job. There are plenty of WordPress developers who do this for 40 hours a week, and if you *really* want to optimize your site, you can get a 100% score on GTmetrix.

But if you only want to spend an hour or two optimizing your site, you can still make huge gains and improvements. If you followed all of the advice in this chapter, you should have seen a pretty big speed improvement through GTmetrix.

In this chapter, you have learned how to monitor performance with GTmetrix, minify static files such as CSS and JavaScript, optimize images, and cache files on your site.

In the next chapter, we're going to configure our theme.

Part 3: Customizing the Appearance and Functionality of Your Store

A fully functional WooCommerce store is a pretty good start. What's even better is a store that looks great and attracts attention. *Part 3* will guide you through the process of enhancing the look and functionality of your store. This section will cover the following chapters:

- *Chapter 10, Setting Up Your Theme*
- *Chapter 11, Customizing the Product Page*
- *Chapter 12, Building a Landing Page*
- *Chapter 13, Creating Plugins for WooCommerce*
- *Chapter 14, Next Steps with WooCommerce*

10
Setting Up Your Theme

Up until this point, we've focused on what your store can do. But just as important as what your store can do is how your store looks. Do you want a vibrant and fun look or a clean and minimalist look? Each will cater to a different audience.

In this chapter, we're going to look into how you can control the important visual aspects of your store. We'll dig into the following:

- Choosing a theme for WooCommerce
- Rearranging the product page
- Adding a product data tab

By the end of this chapter, you should know how to choose a theme built for WooCommerce, customize the appearance of your store so that your store reflects your brand, add a product tab, and install a plugin to see the frontend hooks.

The first thing we're going to explore is choosing a theme for your online store.

Choosing a theme for WooCommerce

There are thousands of themes on `WordPress.org` but not all of these themes are going to work well with WooCommerce. Many of them are designed for brochure websites or blogs. We want a theme that has space for lots of products and ample space on the product page for product details.

We're going to take a look into some of the most popular themes for WooCommerce, including the following:

- Twenty Twenty-Four
- Storefront
- Astra

We're going to look into how to set up each of these themes and the benefits of each one. First up is Twenty Twenty-Four.

Exploring the Twenty Twenty-Four theme

Twenty Twenty-Four is the most recent version of the default theme released with WordPress. Each year, WordPress releases a new default theme and the current version is Twenty Twenty-Four (https://wordpress.org/themes/twentytwentyfour/). All of these themes are available for free and they're a great place to start any WordPress project.

One of the best features of the default themes is they almost always show off the new features built into WordPress. Twenty Twenty-Four includes whole page patterns and template variations so that users can assemble whole pages quickly and easily.

Let's take a look at a real blog post using those new blocks. Follow these steps:

1. Log in to your WordPress admin.

2. Navigate to **Appearance** | **Themes** and you should see **Twenty Twenty-Four** in your list of themes, as we can see in the following screenshot:

Figure 10.1: Finding Twenty Twenty-Four in our installed themes

3. If you don't see Twenty Twenty-Four, then click **Add New Theme** and search for Twenty Twenty-Four to find and install it.

4. Hover over **Twenty Twenty-Four** and click **Activate**.

The theme is now activated. Let's take a look at the theme.

If you browse to the **Shop** page, you'll see a pretty bare page. My shop can be seen in the following screenshot:

Figure 10.2: My site with Twenty Twenty-Four active

The **Shop** page is a bit plain. We need to update the colors so that they're readable. And we're also likely going to want to add more content to this page.

Now, let's edit this page. Follow these steps:

1. In your WordPress admin, click on **Pages**.
2. Find your home page (usually called **Home**).
3. Add or customize the existing blocks.
4. I recommend using the **Cover**, **Best Selling Products**, **Featured Product**, and **On Sale Products** blocks. We already set up some of these blocks in *Chapter 3 – Organizing Products*.

5.   When you're adding or editing blocks, I also recommend using the **Wide** width orientation or **Full** width, which you can see in the following:

Figure 10.3: Try wide and full orientations

6.   If you need to change the color of any fonts, now is the time to do so within the **Block** settings, as seen in the screenshot:

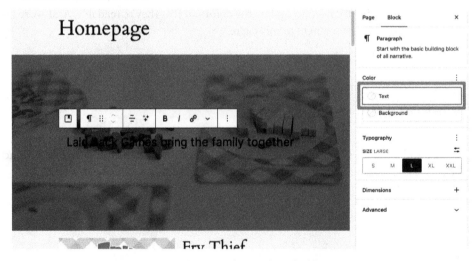

Figure 10.4: Change colors within Block settings

7. I don't love the title **Homepage** on my homepage. It's too obvious and doesn't help visitors. Lucky for us, the Twenty Twenty-Four theme has a special page template that hides the page title. Under **Page** settings, there's a **Template** section. Select **Page No Title**, as seen in this screenshot:

Figure 10.5: Using the Page No Title template

8. Save the page.

Once you've added blocks, using the full width or wide width setting, take a look at the frontend. Here's my very simple home page with the blocks we added earlier with our updated settings and page template:

Figure 10.6: Overview of the home page in Twenty Twenty-Four

Because the theme is so empty, it's perfect for customizing with blocks. If you want to build very customized pages with blocks, then this theme could be a great fit for you.

If you have a more standard e-commerce site and you don't expect to be making tons of custom pages like this, then you might want to check out Storefront and Astra.

Storefront

One of my favorite themes is Storefront (`https://woocommerce.com/storefront/`). It's a free theme built by the WooCommerce team. It's the theme they use to test every feature against, so if you want the most reliable theme or the theme where you know every piece of WooCommerce functionality will look great, then you'll want to check out Storefront.

You may have already installed and activated Storefront by going through the WooCommerce welcome wizard. If you haven't, you can install it by following these steps:

1. Go to your WordPress admin.
2. Go to **Appearance | Themes**.
3. Click **Add New Theme**.
4. Search for `Storefront` and install.

Besides being the official theme for WooCommerce, Storefront has a few things going for it. We're going to look into the following:

- The sticky add-to-cart button
- Paginating between products

Sticky add-to-cart button

One of the most useful features that's included with Storefront is a sticky add-to-cart button. When you're looking at a product page, you'll see the **Add to cart** button in its usual place. This can be seen in the following screenshot:

Figure 10.7: A regular Add to cart button before you scroll down

When you start scrolling down, as soon as you scroll past the **Add to cart** button, it should be added to a sticky bar at the top of the page. Here's what it looks like on my laptop:

Figure 10.8: As you scroll down the page, the Add to cart button sticks to the top of the browser

In our case, since we're looking at a variable product, we see **Select options**. But a simple product would say **Add to cart**. This functionality is built into Storefront. You can disable it by following these steps:

1.   Navigate to the WordPress customizer.

2.   Click on **WooCommerce | Product Page** settings.

3.   Disable the setting for **Sticky Add-To-Cart**, which you can see here:

Figure 10.9: Enabling Sticky Add-To-Cart

4.   Click **Publish** to save your changes.

If for some reason this functionality is turned off by default, you can turn it on in the **Settings** page. I am a big fan of this setting since it really helps users as they scroll down the page. For my store, I'm going to keep this setting on.

Product pagination

Pagination lets you navigate between different areas of the site. While we're on this settings page, let's enable product pagination. This lets us navigate between different products.

You can see what it looks like on my site:

Figure 10.10: Pagination on product pages

In addition, as you hover over the images, they slide out and show the full image and product title:

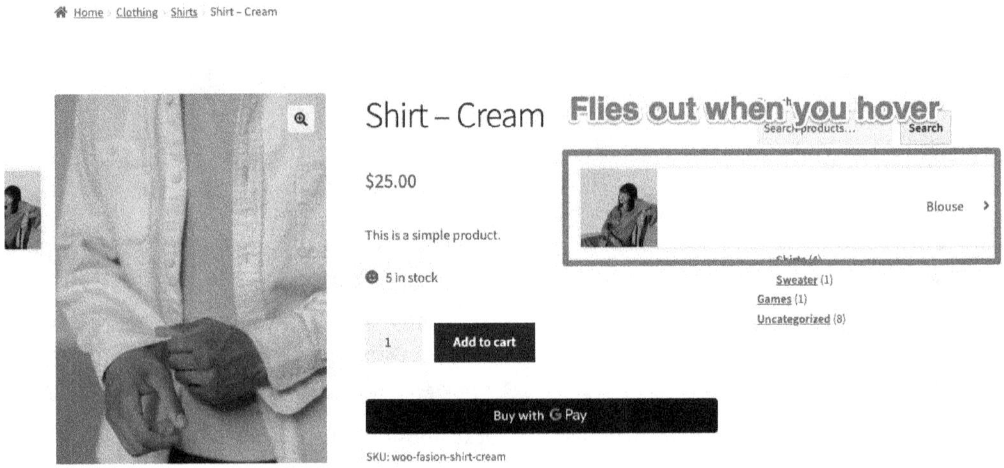

Figure 10.11: Pagination flies out when you hover over it

I'm happy they included this functionality in this theme but for my own purposes, I find it clutters up the product page, and navigating through products this way seems unintuitive. Still, if you like this functionality, go for it. It's great to have it built into the theme instead of having to install a bunch of plugins.

Exploring Astra

There's one final theme that is worth mentioning, this time because it's free and fast. Astra is a very tiny theme and it's designed to load quickly (`https://wordpress.org/themes/astra/`). If you're a fan of speed and lightning-fast pages, then Astra will help you to get there (in addition to the previous chapter).

Here's the GTMetrix report of our site with Storefront:

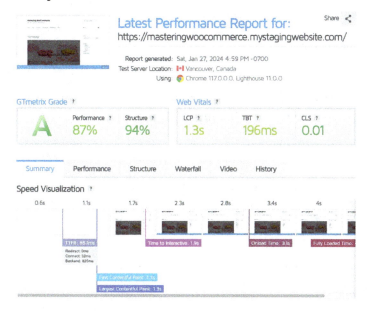

Figure 10.12: A speed test on Storefront

Here's the same report after switching to Astra:

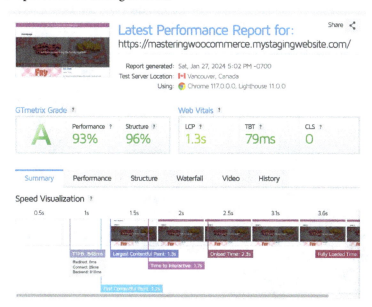

Figure 10.13: A speed test with the Astra theme

Both the **Performance** and **Structure** scores increased into the mid-90s. The **Largest Contentful Paint** (**LCP**) stayed the same, but the **Fully Loaded Time** dropped from four seconds down to three seconds. On a small demo store like this, there isn't much improvement, but if you use your real store, you could see even bigger improvements.

This theme doesn't have many options and, for store owners who value speed, this is immensely helpful.

If you like the idea of a fast theme, you can get Astra for free or you can check out the Pro version of Astra, which has additional features and customization options.

All of the themes we've covered in this section are good at certain things and not great in other areas. It's totally up to you and what's important for your brand when you pick a theme.

Next up, let's look at how we can view the hook on the front of our site so we can customize our theme further.

Rearranging the product page

It's very common for store owners to want to customize the frontend of their store. They might really like the look of the product page, but they want to rearrange certain elements.

In this section, we're going to install a plugin to help us understand how the WooCommerce code works so that we can modify it.

As an example, let's say we take a look at our product page, and we don't think listing the category is important for our store. We could move that lower down the page. But how would we even begin to do that?

In the following screenshot, how do we move the product price lower down the page? Take a look:

Figure 10.14: I'd like to move the price lower down the page

The answer is simple but not easy. The first thing we have to do is find where this code is coming from and how it's being displayed on the page. We could go through thousands of lines of code in WooCommerce or we could use what I call hook visualizer tools, which we will be looking at in the next section.

Installing hook visualizers

Hook visualizers are tools made for developers. They help you to know exactly what code is running and when. Let's install one and you'll see how useful they can be.

I'm going to install my new favorite hook visualizer, which has some great features that we'll show next. But there is a much more popular hook visualizer called **WP Hooks Finder**, which you could install. Let's install **Another Show Hooks**. Follow these steps:

1. In your WordPress admin, go to **Plugins | Add New**. Search for Another Show Hooks:

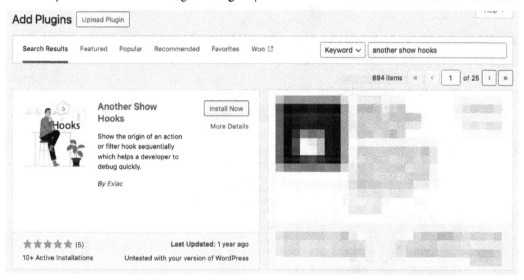

Figure 10.15: Installing the Another Show Hooks plugin

2. Install and activate the plugin.

3.   Go to the frontend of your site and you'll see a new menu:

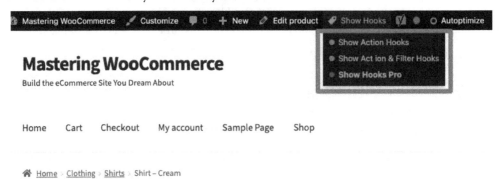

Figure 10.16: Show Action Hooks

4.   Hover over **Show Hooks** and click **Show Action Hooks**. This will make the page reload with a ton of extra data, which we can see in the following:

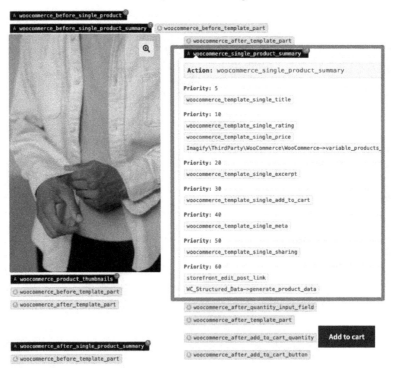

Figure 10.17: Hover over action hooks to see what code is executing and when

Here, you can see every action. Actions in WordPress are where code is executed. In *Figure 10.17*, under **woocommerce_single_product_summary**, you can see there are ten separate pieces of functionality. We can see the priority for each of them, which is the order their code is executed.

Once we know this information, we can write code to change that priority.

If you're familiar with WordPress actions (`https://developer.wordpress.org/plugins/hooks/actions/`), you can unhook any code and write new code to hook it in a new place. This isn't a book about the WordPress action system, so we can't go into too much detail here. But this is the real power of WooCommerce. With the WordPress action system, you can remove or add anything you want.

Stop showing hooks

The Simply Show Hooks plugin only shows hooks to the admin of the site. So, your users won't see these. However, they still take up a lot of space. At any point, you can hover over **Simply Show Hooks** and click **Stop Showing Action Hooks**.

Browsing through code for actions

I'm a huge fan of hook visualizers because you can navigate to a page, see the actions on that page, and then reverse-engineer what you have to do. But sometimes, you don't even know where to start. In that case, it's helpful to be familiar with the code base.

WooCommerce has a helpful beginner guide to actions (`https://woocommerce.com/document/introduction-to-hooks-actions-and-filters/`). More importantly, they also have a complete hook reference (`https://woocommerce.github.io/code-reference/hooks/hooks.html`). You can browse through this resource and find any hook and read more about it.

As an example, I can search for `woocommerce_single_product_summary`. This can be seen in the following screenshot:

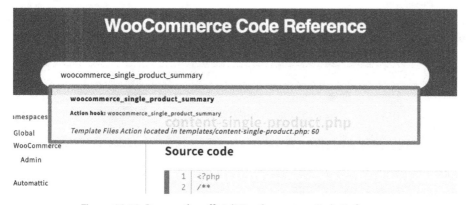

Figure 10.18: Browse the official WooCommerce Code Reference

I can click **woocommerce_single_product_summary** and I will be taken to the exact code that controls this functionality. We can see that in the following screenshot:

```
46          <div class="summary entry-summary">
47              <?php
48              /**
49               * Hook: woocommerce_single_product_summary.
50               *
51               * @hooked woocommerce_template_single_title - 5
52               * @hooked woocommerce_template_single_rating - 10
53               * @hooked woocommerce_template_single_price - 10
54               * @hooked woocommerce_template_single_excerpt - 20
55               * @hooked woocommerce_template_single_add_to_cart - 30
56               * @hooked woocommerce_template_single_meta - 40
57               * @hooked woocommerce_template_single_sharing - 50
58               * @hooked WC_Structured_Data::generate_product_data() - 60
59               */
60              do_action( 'woocommerce_single_product_summary' );
61              ?>
62          </div>
```

Figure 10.19: Browse WooCommerce code directly in the WooCommerce Code Reference

If you are comfortable with code, then browsing through the code might be the best option for you. But for most of us, I recommend starting with a hook visualizer tool. We can see most of what we need with that tool.

Demo – move the product price

In *Chapter 13*, we'll create a proper plugin to customize WooCommerce code. But let's write a simple customization now that we know how to find the right hooks within WooCommerce.

I created a custom plugin for my site. I wrote the following code in a code editor and uploaded it to my site. You'll want to upload it to wp-content/plugins under your site. Then, activate the plugin:

```
<?php
/*
Plugin Name: WooCommerce Move Product Page Price Lower
Plugin URI:  https://gist.github.com/BFTrick/
e19bd2bb648558ca78279c3d5ba9524e
Description: A demo plugin to move the WooCommerce simple product
price lower
Version:     1.0
Author:      Patrick Rauland
Author URI:  http://speakinginbytes.com
```

```
License:       GPL2
License URI: https://www.gnu.org/licenses/gpl-2.0.html
Text Domain: woocommerce-move-product-price
Domain Path: /languages
*/


// We're going to run our code after WooCommerce is loaded
add_action( 'woocommerce_loaded', 'mastering_woocommerce_lower_
product_price', 20 );


// For our simple product's we're going to move the product price
lower down the page.
// May be theme dependent. This code was tested against Storefront
function mastering_woocommerce_lower_product_price( ) {
    remove_action( 'woocommerce_single_product_summary', 'woocommerce_
template_single_price', 10 );
    add_action( 'woocommerce_single_product_summary', 'woocommerce_
template_single_price', 35 );
}
```

And here's what this looks like on the frontend:

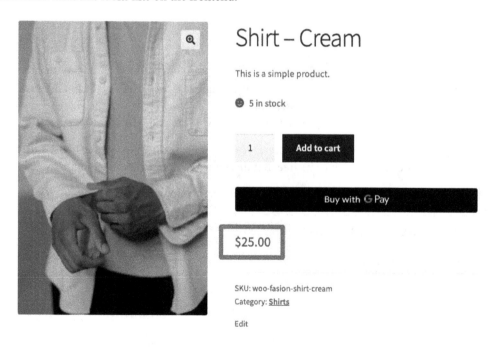

Figure 10.20: We successfully moved the product price lower

I used the hook visualizer to see when the product price code was executed. For my site, it was `woocommerce_single_product_summary` at priority 10. I used the `remove_action` WordPress function to remove that code. You can see that essential line here:

```
remove_action( 'woocommerce_single_product_summary', 'woocommerce_
template_single_price', 10 );
```

I again used the hook visualizer to see at what point other code is executing and I added my price *after* the **Add to cart** button. For this site, I used `add_action` to add the code to the same action, `woocommerce_single_product_summary`, but now at priority 35. Here's the essential line that re-adds the functionality:

```
add_action( 'woocommerce_single_product_summary', 'woocommerce_
template_single_price', 35 );
```

If you want to dig into how exactly this code works. I recommend the WordPress documentation about the action system.

It explains how to remove actions and add actions somewhere else: `https://developer.wordpress.org/plugins/hooks/actions/`

Now that we know how to move a few things around on our page, let's look into how to add a product data tab.

Adding a product data tab

WooCommerce has a nice tab system on the product page. It's the perfect place to add custom information to your product. If you have extra information that you really want to share with your audience rather than burying it in the product description, you can add it to a custom tab.

This can be done with code but there are also several easy-to-use plugins. We're going to use **Custom Product** tabs for WooCommerce since it's free. But there are more powerful paid plugins available.

We're going to install a custom tab plugin and then configure that plugin.

Installing a custom tab plugin

Let's start by installing the right plugin. Follow these steps:

1. In your WordPress admin, go to **Plugins** | **Add New**.
2. Search for `Custom Product Tabs for WooCommerce`, as seen here:

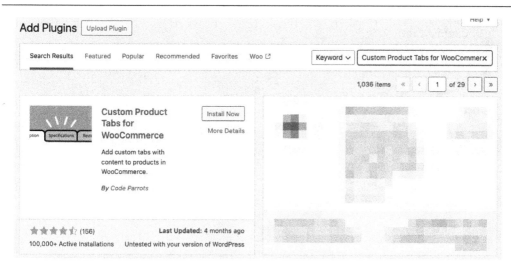

Figure 10.21: Installing Custom Product Tabs for WooCommerce

3. Install and activate the plugin.

Adding a custom tab

Now that we have the plugin installed, we can add a custom tab to a page. Follow these steps:

1. In your WordPress admin, edit one of your products.

2. If you scroll down to the **Product data** panel, you'll see an extra tab called **Custom Tabs**, as seen here:

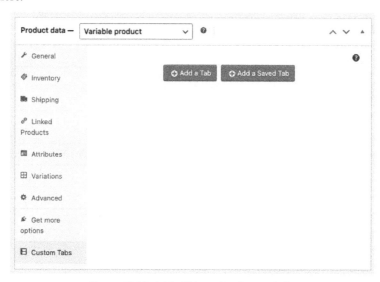

Figure 10.22: Add a Tab under Custom Tabs

3. Click **Add a Tab**.

4. Add a title and description for the tab, as shown here:

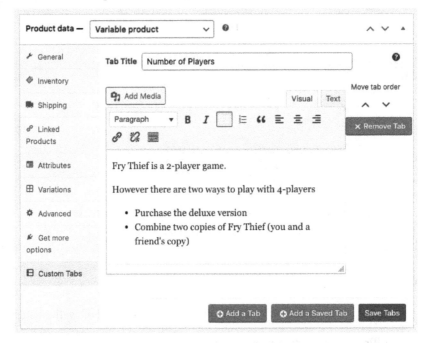

Figure 10.23: Add tab content and then click Save Tabs

5. Click **Save Tabs**.

6. Click **Update** to finish updating the product.

Now, if you take a look at the frontend, you'll see our custom tab and content:

Figure 10.24: Our custom tab on the frontend

This is only the start of custom tabs. You can add additional tabs, and you can even save tabs and then apply them to multiple or even all of your product pages. This is really helpful for some information, such as sizing charts, assembly instructions, and more.

Extensive customizations using child themes

We talked a lot about how you can customize your theme using the customizer, using hooks, and installing additional plugins. There's one more common way to customize your theme that's really useful for sites that want to customize every aspect of their store. That's using a **child theme** (https://developer.wordpress.org/themes/advanced-topics/child-themes/).

I like to think of a child theme as a second draft. You choose a theme to be the parent and then you modify from there. You could choose any WordPress theme as the parent, and I would suggest starting with one of the three themes we already covered in this chapter.

Then, in the child theme, you can customize the styling (CSS) and the functionality (the hooks).

One of the most important benefits of a child theme is that it allows you to update the parent theme and bring in new changes while still keeping your modifications intact.

An additional benefit is that you don't need to create a special plugin like we did in the Rearranging the product page section. Instead, you can paste any PHP code into the `functions.php` file in your theme.

Creating a child theme is beneficial if you want to invest in your theme. But not all businesses want to spend time curating the look and feel of their site. They can often accomplish what they need in the customizer without needing to create a child theme. Hence why I showed off that functionality in this chapter. But for those of you who are building for the long term and have the ability to write code, creating a child theme is the right strategy.

Summary

Customizing the frontend of your store is important. It's what people see and remember, and there are lots of ways to do it. By now you should be familiar with three of the most popular themes for WooCommerce and you hopefully installed one of these themes. You also know how to look at the front of your site with a hook visualizer so you can see where certain code actions are taking place. With plugins, you can customize the tabs that appear on the product pages. You now know how to customize your own store after reading this chapter.

In the next chapter, we're going to look into customizing specific parts of the product page with functionalities you might not have considered.

11
Customizing the Product Page

We've set up our theme. Now, it's time to add more information to our product page in order to make it enticing and pave the way for users to buy our products.

We're going to add social proof and a few different types of media to the product page. Social proof shows users that other people are actively using and purchasing products on this site, which makes our site feel less risky, and adding extra media such as videos or 360-degree images will give the user more context to help them determine whether this is the right product for them.

The following topics will be covered in this chapter:

- Adding social proof
- Adding a video tab to the product page
- Adding 360-degree images

By the end of this chapter, you should know how to make your product page a lot more informative and visually interesting.

Adding social proof (FOMO)

Human beings are social creatures. If we see another person doing something, then we assume that it's safe and maybe even beneficial for us to do that same thing. That's why, when we see a whole group of people gathering, we want to see what they're gathering around. We call this behavior social proof. And it's something that we can also add to our website.

There are lots of types of social proof, including the following:

- Product reviews
- Followers on social media

But in this section, we're going to talk about seeing other people doing the thing that you want them to do. If you visit the Fomo website (https://fomo.com), you'll see the activity at the bottom of the site. You can see this in the following screenshot:

Figure 11.1: FOMO in use on a WooCommerce site

We can see that a real user was active on the site. That makes it much more likely for me to do that same activity.

We're going to set up FOMO on our e-commerce site so that anytime someone purchases a product, we broadcast that to other users. The goal is to make all the other users more likely to do the same activity.

After setting up FOMO, we'll customize the notification.

Setting up FOMO

The first step is to go to https://fomo.com and create an account. Once you have done that, you can connect it to your store. Follow these steps to set up FOMO:

1.   Log in to your FOMO account.

2. Add your first site, by adding your store's URL, as shown in the following screenshot:

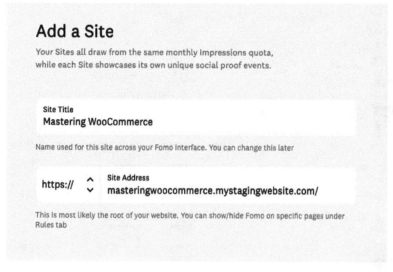

Figure 11.2: Add your store's URL to FOMO

3. Next, add a new type of notification. Click on **Add Notifications**.

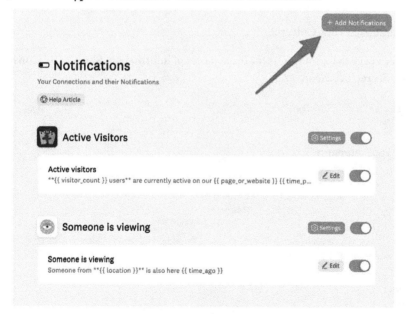

Figure 11.3: Add a Notification

4.   On the next page, you'll see a ton of integrations. Choose **WooCommerce**.

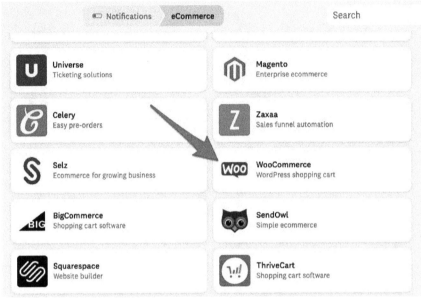

Figure 11.4: Add WooCommerce notification

5.   On the next page, click **Connect**.

Now, we land on a page that prompts us for three pieces of information: **Shop URL**, **Consumer Key**, and **Consumer Secret**. See *Figure 12.5*.

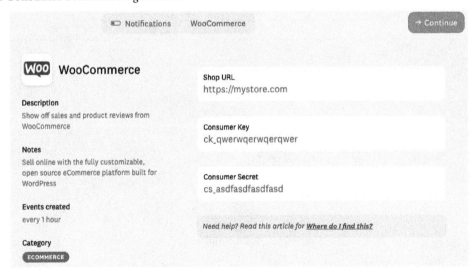

Figure 11.5: Fields we need to connect WooCommerce to FOMO

1. We've seen this before, in *Chapter 10 – Setting Up Your Theme*. We're connecting our store to FOMO via an API. Go ahead and enter your store URL here, and then we'll create an API connection in WooCommerce.

2. In WooCommerce, go to **Settings** | **Advanced** | **REST API** and then click on **Add key**. See *Figure 11.6*.

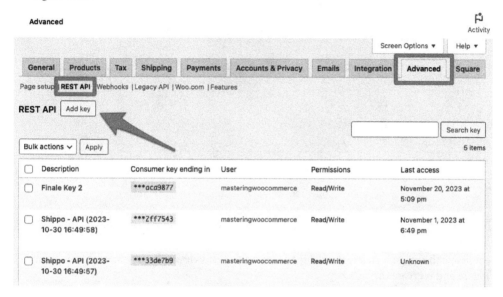

Figure 11.6: Add REST API key

3. On the next page, we can configure a few details about the API key. We want to add a descriptor of FOMO so we know what the API key is for. And we want to give FOMO Read/Write permissions. See *Figure 11.7*.

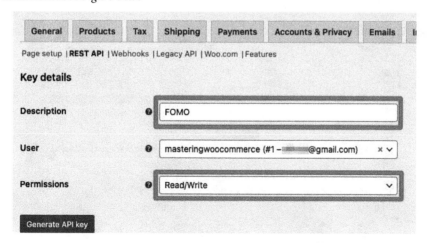

Figure 11.7: Set permissions for the REST API key

4. Copy your consumer key and consumer secret into FOMO. See *Figure 12.8*.

Figure 11.8: Copy Consumer key and Consumer secret

> **Warning**
>
> If you leave this page, you won't be able to come back and view this information. You'll have to delete and recreate the API key.

5. Finally, we can move on to the next step in FOMO. As shown in the following screenshot, after filling in your information, click **Continue**.

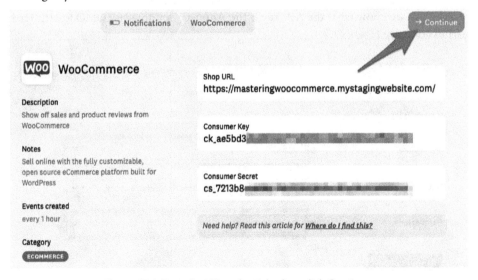

Figure 11.9: Paste in API credentials, then click Continue

That created our notifications in FOMO. That's a good chunk of the work done. You can see all of your notifications on the **Notifications** page in FOMO. See *Figure 11.10*.

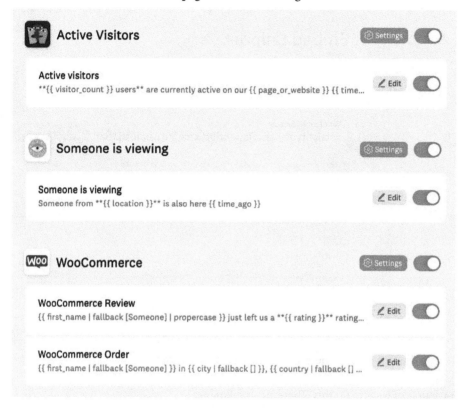

Figure 11.10: Reviewing notifications in FOMO

1. From here, you can disable or edit these notifications.

    The notifications are set up in FOMO. The service listens for orders and reviews. As it picks up that data, it creates notifications that we can display on our site.

2. The final step is adding JavaScript code so users will actually see these notifications. See *Figure 11.11*.

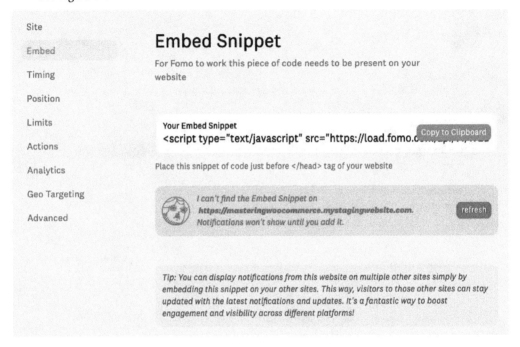

Figure 11.11: Embedding FOMO script

There are a couple of ways to do this. These are as follows:

- We could add the JavaScript code directly to our theme files. If we custom-created our own theme or child theme, we could take this route.

- We could add the JavaScript code through a plugin. This is a bit easier, and anyone can do it.

Let's add that JavaScript code via a plugin. Perform the following steps:

1. In your WordPress admin, install and activate the **WPCode – Insert Headers and Footers** plugin (`https://wordpress.org/plugins/insert-headers-and-footers/`).

2. Copy **Your Embed Snippet** from FOMO under **Settings > Embed**.

3. In WordPress, navigate to **Code Snippets | Header & Footer** and paste in the FOMO embed code. See *Figure 11.12*.

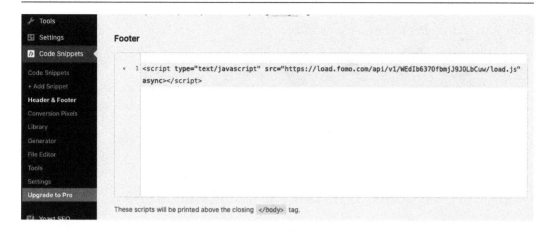

Figure 11.12: Adding custom Javascript

4. Then click **Save Changes** at the top of the page.

5. The code is now on your site. Let's make sure that FOMO can detect the script. If you go back to the **FOMO Embed Settings** page, you'll see a success message showing the script is detectable by FOMO and we should have notifications on our site. Refer to *Figure 11.13*.

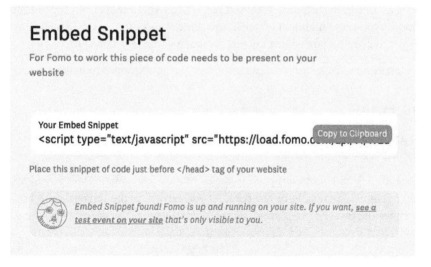

Figure 11.13: Verifying script is detected in FOMO

You can even click the link and see a test notification on your site. See the following screenshot.

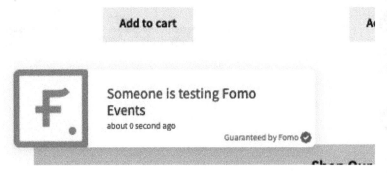

Figure 11.14: Verifying FOMO on your store

And now we have FOMO set up, and we've verified it works. You're ready to start using the FOMO plugin. However, you also have the option to customize further.

Customizing notifications

For the most part, FOMO will automatically pull in useful notifications and display them on your site. However, you can always customize what you see. If you want to show really big product photos, you can do that, or you could highlight where in the world your customers are purchasing from, or you could keep it really minimal and just say that someone purchased a product.

In the end, it's totally up to you, and you can customize your notifications to make your brand.

Removing events

If you're testing products, orders, payments, and similar features on your site, then you probably have some test data that you want to clear up.

In FOMO, you can click on **Events** and you can see all of the events that FOMO picked up. You can click **Remove** to remove any invalid or test data. As you can see in the screenshot in *Figure 11.15*, Next, I will remove the two test purchases. I'd recommend you do the same, to keep your records clean and accurate.

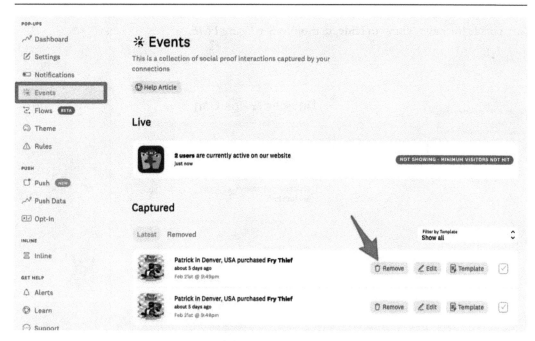

Figure 11.15: Removing test notifications

Once we've removed any incorrect data, we can customize on what devices these notifications are displayed.

Hiding notifications on mobile

Sometimes, social proof such as FOMO sounds great, but when you're on a tiny device, it might actually prevent you from viewing the site entirely.

If you want to give your users a better mobile experience, you can disable FOMO on mobile devices.

Perform the following steps:

1. In FOMO, navigate to **Settings** | **Position**.

2. Enable the radio button to **Hide on mobile**. See *Figure 11.16*.

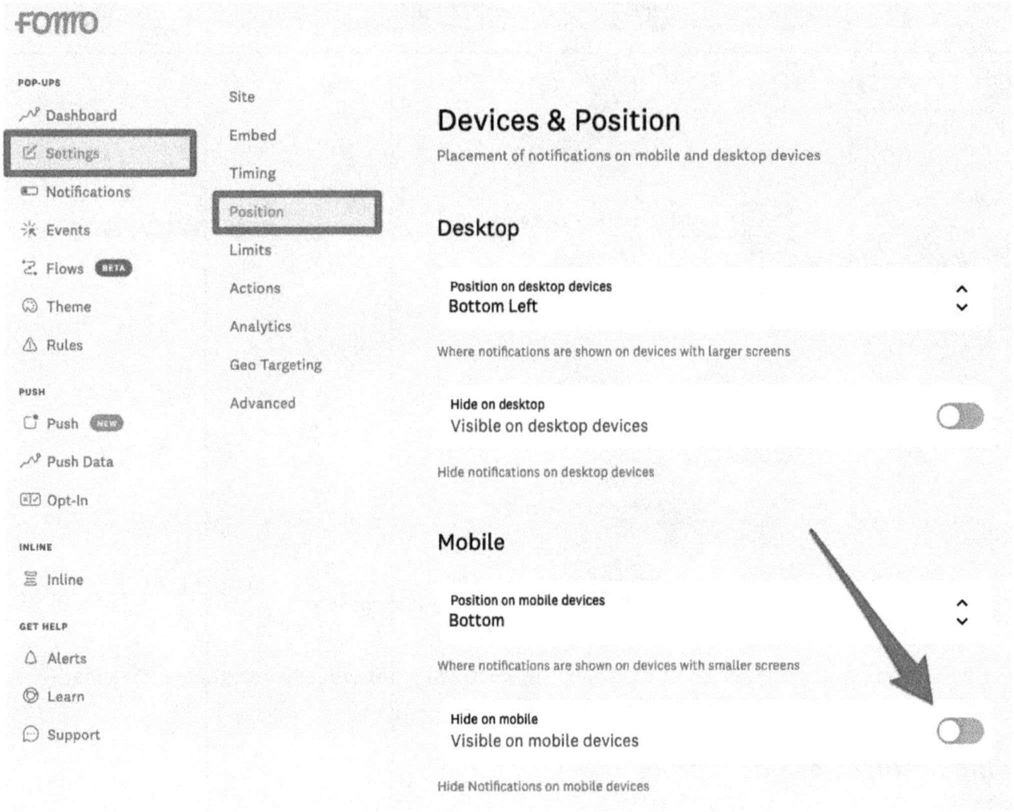

Figure 11.16: Disabling FOMO on mobile

3. You can now click **Save**.

FOMO will now only display on tablets and on desktop devices. It won't display on mobile devices.

Customizing notification frequency

You might also want to choose how often these notifications appear. For busy stores, you might want to show a notification every 2 seconds. But for a smaller, more niche store, maybe you only want to show a notification every 12 seconds.

As an example, I might only want to show notifications to users who have been on my site for 10 seconds or more.

To do so, perform the following steps:

1. In FOMO, navigate to **Settings | Timing**.

2. Change **Initial delay (seconds)** from 3 to 10, as shown in the upcoming screenshot.

3. Change **Time between** to 20. See *Figure 11.17*.

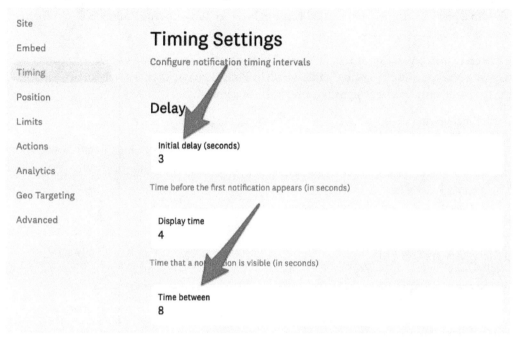

Figure 11.17: Timing settings in FOMO

4. You can now click **Save**.

As you can see, FOMO is pretty much a set-it-and-forget-it technology. There isn't much management, and it shows new visitors to your store what other people are actually looking at and buying.

Now that users know that other people are actively buying products from our store, let's look at providing users with video information on the product page.

Adding a video tab

People like to consume information in a variety of ways. Some people like reading text and images, others like word of mouth and talking about a product, and others still prefer a video that's fully engaging for a few minutes and describes the product in detail. Most of us like some combination of these at different times for different products.

In this section, we're going to add a **video** tab to our product page so that there's a designated place for it – out of the way but easy to navigate.

We're going to start by installing the video tab and then move on to customizing it.

Installing a video tab

There are a number of plugins that let you add videos to your WooCommerce page, many of which are free. In a previous chapter, we installed **Custom Product Tabs for WooCommerce**. You are free to reuse that plugin.

I've chosen to install WooCommerce Product Tabs (`https://wordpress.org/plugins/woocommerce-product-tabs/`). This is built by Barn2 Plugins. They're a well-known company in the space and the plugin is designed in a very simple manner.

Perform the following steps:

1. Search for **WooCommerce Product Tabs** on your WordPress site. Click **Install Now** on the plugin from Barn 2 Plugins.

2. Then click **Activate**.

And that's all we need to do to install it – pretty fast and easy.

Adding an extra tab

After you activate the plugin, as shown in *Figure 11.18,* you should see a welcome wizard.

Figure 11.18: Wizard for WooCommerce Product Tabs

You can, of course, skip this wizard. However, let's go through this as it will definitely help speed things up. Let's set up a tab:

1. Click **Next**.

2. Now we can add a tab title and tab content to this tab. See *Figure 11.19.*

Figure 11.19: Set Tab content

3. Click **Create**.

4. On the next page, they'll ask if you want to install some of their other plugins. We don't need these for our store so I'll skip over that and click **Finish Setup**.

After this, we'll see a confirmation page with a few options.

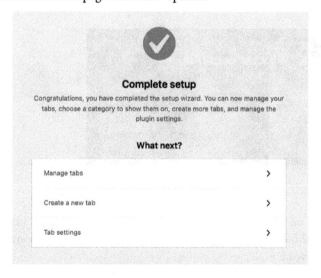

Figure 11.20: Completing WooCommerce Product Tabs wizard

You can, of course, click into any of these settings. But you can also manage your tabs in WordPress admin, under **Products | Product Tabs**.

There is one setting that many people reading this book will want to change. And that's the priority. That controls where this tab appears within the rest of the tabs. Here's how to change the priority of your tabs:

1. Go to the **Manage Tabs** page.

2. Click on the tab you created in the welcome wizard. Or, you can create a new one.

3. In your tab, set **Priority** to 50 or some high number.

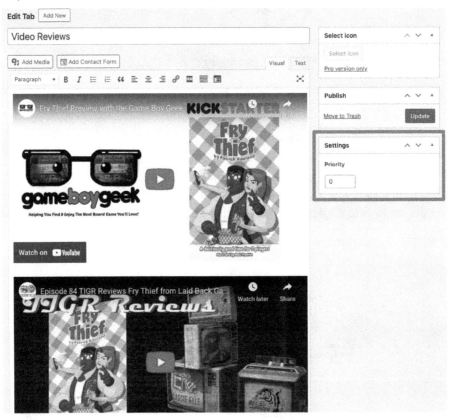

Figure 11.21: Set the priority of the custom tab

4. And while we're on this page, let's set one other common setting. Let's change the visibility so this tab only shows up for a particular category. Scroll down and choose **Show on specific categories**. See *Figure 11.22.*

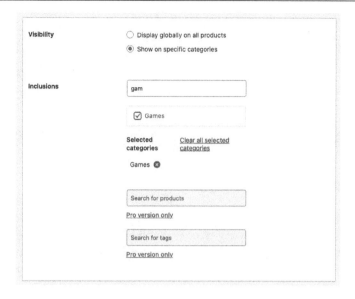

Figure 11.22: Display tab on specified categories

5. Then you can search for your category and select it.

6. At the end, click **Update** to save your changes.

On the frontend of one of our games, we now have a new tab, as seen in *Figure 11.23*.

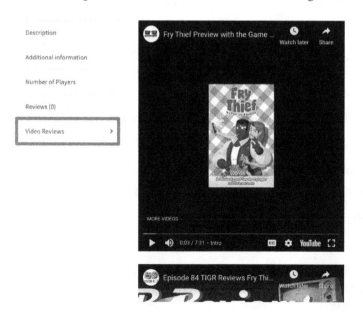

Figure 11.23: Viewing the custom tab on frontend

As mentioned earlier, you can apply this to a whole category, or across your whole site. This could be very useful for FAQ content.

If you want to have a unique tab for each product, then you may want to get the Pro version of this plugin. Or, you can use another plugin, such as **Custom Product Tabs for WooCommerce**, highlighted earlier in this book.

We've added our unique tab. Now let's add 360-degree images to make our product pages even more appealing.

Displaying 360-degree images

Humans are very visual. We like to see images alongside written descriptions, and one image isn't enough. Customers want to see multiple angles. Several well-known online stores (`https://cxl.com/blog/how-images-can-boost-your-conversion-rate/`) have tested this, and they've seen conversions improve by 10-30%. That's why we're going to look into displaying images that cover 360 degrees of a product.

We're going to use WooCommerce 360° Image (`https://woocommerce.com/products/woocommerce-360-image/`) to show off multiple angles of our products. We're going to start by installing the plugin, followed by adding product photos. When we're done, users will be able to see every aspect of our product and will be much more likely to add the item to their cart and check out.

Installing WooCommerce 360° Image

This is a premium plugin, so if you want to use it, you'll have to purchase it from `woocommerce.com`.

Alternatively, you can try a free plugin from WordPress.org. I haven't personally tried this plugin – `https://wordpress.org/plugins/360-product-viewer-for-woocommerce/` – and it only has one review, so user beware!

Perform the following steps to install WooCommerce 360° Image:

1. Purchase WooCommerce 360° Image on `woocommerce.com`.
2. Download the ZIP file with the plugin.
3. On your WordPress site, go to **Plugins | Add New** and upload the ZIP file.
4. Activate the plugin after it has finished uploading.

That's all we need from an installation standpoint. Now, we need to configure it with our products.

If you don't have your own photos, the documentation for WooCommerce 360° Image (`https://woocommerce.com/document/woocommerce-360-image/`) links to a ZIP file that contains a few example images.

Adding 360-degree images to products

Let's create a new product, and then add images to that product. Perform the following steps:

1. In admin, create a new product by clicking on **Products | Add New**.

2. Give the product a title, a price, and a description, if you so wish.

3. Check the **Replace Image with 360 Image** box:

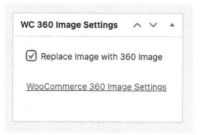

Figure 11.24: Replace the main product image with a 360-degree image

4. Under **Product gallery**, click **Add product gallery images** to upload your images, as shown in the following screenshot:

Figure 12.25: Adding 360-degree images to product gallery

5.   Publish the product so it's live on the site.

On the frontend, you'll see the 360-degree image functionality. You can rotate the product left and right. You can also click the play icon and watch the product spin around, as shown in the following screenshot:

Figure 11.26: 360-degree image on frontend

Until we get to the point of shopping through virtual reality, this is the closest thing for store owners. From a technical perspective, it's easy to set up. The hardest part is getting good-quality pictures and getting enough photos.

I've implemented 3D models at several companies but 3D modeling technology is just a little too expensive for most online stores. You can get 80% of the benefit of a 3D model with 360-degree photos and this extension.

Summary

A default WooCommerce product page is fine. It does the job, but it doesn't stand out. The technologies that we covered in this chapter make your product page stand out and make it more likely that someone will purchase from you.

By now, you should be able to integrate your store with FOMO and display social proof to new users. You can add a **video** tab to your product pages and display multiple videos. And you can use a 360-degree image to show multiple angles of a product.

All of these things will make your store more appealing, and they require almost no maintenance or ongoing work.

In the next chapter, we will look at how to build a landing page with WooCommerce Blocks.

12

Building a Landing Page

A common strategy when announcing a new product or service is to create a landing page for that product. These can be used to educate potential customers about the new product or to get sales. They're usually quite in-depth with images, text, headlines, quotes, and user-generated content (also called UGC). These all help a potential customer to understand what the product is, and why they would want it.

Landing pages are also designed so that all of the information is on one page. That way, users don't get lost, and they hopefully sign up with their email or make a purchase. Once you know how to build a landing page, you can build one quickly to gauge interest in a potential product before launching, and once you have a product, you can secure sales.

We're going to build a landing page to capture pre-orders for a new product that we're launching. Then, we are going to start by building a regular landing page, after which, we will add e-commerce functionality to the landing page. Finally, we're going to create multiple versions of the landing page and use A/B testing software to figure out which version will lead to better outcomes (email sign-ups or sales).

Throughout this chapter, I'm going to create a landing page for a new board game (Broken and Beautiful), but you can create a landing page for any product.

The following topics will be covered in this chapter:

- Building a long-form landing page
- Adding e-commerce features to a landing page
- Measure and test everything

By the end of this chapter, you should be able to build your own e-commerce page, add the ability to purchase on that page and be able to measure what visitors do on that page.

Building a long-form landing page

Landing pages are typically a special type of page that is typically used to sell a product or service, and they come in all sorts of formats. Sometimes, they're very short with just an email, a sign-up form, and a photo. We're going to build a long-form landing page for a new product that we're offering.

This longer landing page gives us plenty of space to explain the product, see customer feedback, and see photos and videos.

Let's start by creating a page, then we'll structure our page so that we know where to add the content, and finally, we'll add the content to our landing page.

Creating a new page

Before we can build a landing page to sell someone a product, we must start with the basics by creating a page in WordPress. Let's build a solid foundation before we get into the more technical part.

Create a new page by following these steps:

1. In your WordPress admin center, go to **Pages**.

2. Click on **Add New**, which you can see here:

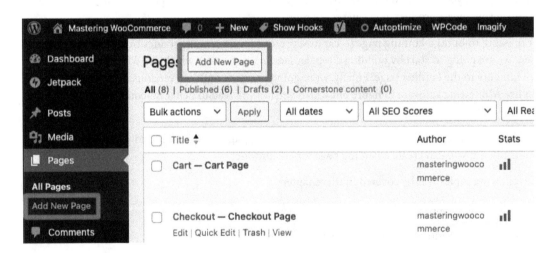

Figure 12.1: Adding a new page

3. Add a title to the new page, which you can see here:

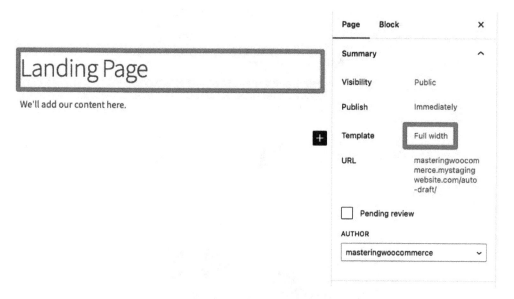

Figure 12.2: Adding a title to our page

And that's all we have to do. There is one optional setting that you can change, which will make a big difference to the landing pages: enabling a full-width template. Landing pages should be focused exclusively on selling the product or the service. You don't want, or need, a sidebar to distract your visitors from other information.

In my Storefront theme, I have an option to remove the sidebar, as you can see in *Figure 12.2*. Many themes have similar template functionality. If you don't, you can reach out to your theme developer, or browse through their documentation.

> **Look for a full-width template**
>
> Many, but not all, themes have an option for removing a sidebar on a page. If you really want to use landing pages, you should look into a theme that supports this option.

Understanding the structure of a landing page

To get an idea of what a landing page looks like, let's take a look at a large WooCommerce store called Xero Shoes (https://xeroshoes.com/). They create a landing page for their twice-annual product launches. This is their newest Spring 2024 product launch (https://xeroshoes.com/spring24sale/) that educates their fanbase about all of the new styles coming out. It's very long and there isn't enough space to show you the whole thing, so I encourage you to visit the page to get the full experience. See *Figure 12.3*:

Figure 12.3: A real landing page for a WooCommerce store

On the landing page, as shown in *Figure 12.3* you can see that there are lots of individual sections. They highlight the following:

- The headline
- Countdown clock
- Gallery of new products
- Typically a call to action, or in this case a call to action for each product
- Reviews or testimonials (below our screenshot)

Let's add these elements to our page.

Add content to a landing page

Now that we have an empty page ready to go, we need to add some content to it. With WordPress's new block interface, there's a lot of potential content that we can add. Let's start by adding a few different types of content to our page. You can add any content you want to your page, but I'm going to focus on visual content, to draw our reader's eye to the page.

Let's start with a cover image, which we can do with a few simple steps:

1. On the **Edit** page screen for your landing page, click the + button to add a new block.
2. Add a **Cover** block, as shown in the following screenshot:

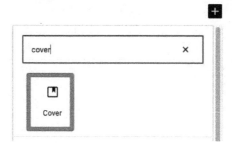

Figure 12.4: Adding a cover block

3. Upload an image that's a visual representation of the product.
4. Add a headline by adding a **Heading** block, similar to the previous screenshot.

5. If you're using Astra, you can check **Disable Title** under **Astra Settings** in the sidebar to hide the default title. In many cases, this is fine, as long as we describe the page in the cover image.

Here's what that looks like on the **Edit** page:

Figure 12.5: A cover block in the admin

And here's what it looks like on the frontend:

Figure 12.6: The start of our landing page

A strong image with a headline is a great start to building an effective landing page. Let's look into adding some features to the page. Follow these steps:

1. On the **Edit** page, add a **New Paragraph** block, and add a short description of the game.

2. Then add a **Pull Quote** block and add the quote.

3. Add a **Heading** block.

4. Add **Media** and **Text** blocks. As you add these blocks and fill them with content, feel free to alternate where the images are. It's very common to alternate images from left to right.

5. Add a **Video** block and add a bit more information about your product.

6. Add a **Quote** block with some social proof that your product is actually good.

7. Save/update your page.

And it should look something like what is shown in *Figure 12.7*:

Figure 12.7: A filled-out landing page

The new blocks in WordPress are incredibly powerful for creating landing pages. I created my landing page in less than 10 minutes with a handful of images. If you spend the time to craft a well-honed marketing message and spend a little more time tweaking the page design itself, you can create a gorgeous landing page for your store!

Now that we have an attractive landing page that clearly explains our product and entices users to buy, we can now look at how to add e-commerce functionality to a landing page.

Adding e-commerce to a landing page

Our landing page is looking pretty good. Now that we've done a good job of showing off our awesome product with different types of content, we just need to give viewers a way to purchase the product. There are a couple of ways to do this: we can use the **Featured Product** block, we can add an **Add to cart** button to the page, or we could also use the **One Page Checkout** extension.

Let's start with the easiest method, which is adding a featured product block. Then we'll look into adding an **Add to cart** button directly to the page.

Adding a featured product

Before we can let users add a product to their cart, we have to add the product to our store. We don't have to add every single field, but we need to add enough to display useful information on the landing page.

Let's add the product first. Follow these steps:

1. Under **Products**, navigate to **Add New** as in the following screenshot:

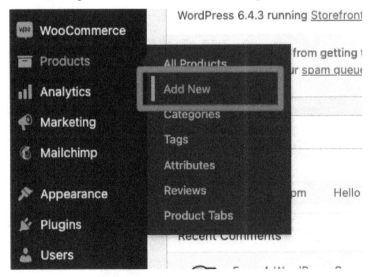

Figure 12.8: Adding a new product

We're creating a simple product for this example. But you can also use any of the products we created in *Chapter 2, All About Configuring Products*.

1. We need to fill in the following fields for our products:

    -   **Add a title**
    -   **Add a product image**
    -   **Add a description**
    -   **Set a price**

2. **Save** the product.

Now, on our landing page, we can add a **Featured Product** block. To add one, follow these steps:

1. Add a new **Featured Product** block.
2. Search for the product that we just added, as shown in the following screenshot:

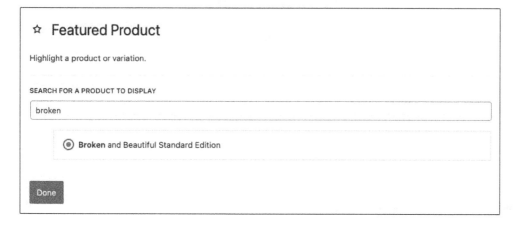

Figure 12.9: Adding a Featured Product block

3. You can select the featured product to tweak the overlay and button colors, and it should look something like this on the frontend:

Figure 12.10: A featured product on the frontend

This certainly works, and it is a nice way to break up large sections of text. You can add a few different call-to-actions throughout the page, such as a button. As soon as someone is convinced that this is the right product for them, they'll click the button and go to the product page, where they can add the product to their cart.

Adding an add-to-cart button

If you want to take what we've already done, and just have a button instead of a whole **Featured Product** block, you can do so by creating an add-to-cart button.

We can do this by creating a magic URL that automatically adds a product to the cart and loads the cart page. It will look something like this: `https://yourstore.com/cart/?add-to-cart={{product-id}}`

You just need to replace the dummy domain with your domain and add the product ID to the URL. Let's start by finding the product ID.

Finding the product ID

We can find a product ID by going to our list of products in our WordPress admin. If you hover over the product, you'll be able to see an ID number. This is shown in the following image:

Figure 12.11: Finding a product ID

Next up, we can add our button.

Adding the button

Once we have our ID, we can add our button. On the **Edit Page** screen for our landing page, do the following:

1. Add a new block. In our case, the block that we are adding is **Button**.

2. Select the **Button** block.

3. Enter text for the button. I suggest Add to cart, followed by the price.

4. For the link, add https://yourstore.com/cart/?add-to-cart={{product-id}}. Of course, replace the domain with your own domain and add your product ID.

5. Then, you can customize the settings and styles for the button as shown in the following screenshot. I made mine wider and changed the colors to dark red and white.

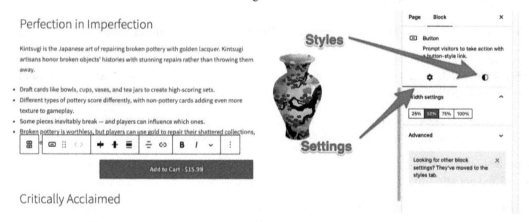

Figure 12.12: Customizing an Add to Cart button

6. Save your page.

If anyone clicks the button they should land on the cart page with the product already in their cart. This can be seen in the following screenshot, which has been taken from my site:

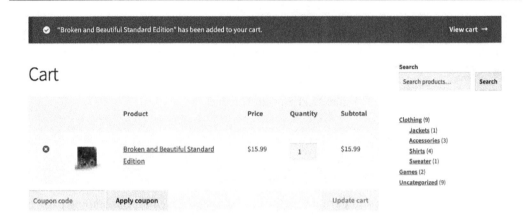

Figure 12.13: Adding a product to the cart from the landing page

We now have a landing page that promotes the product and the user can add it to the cart straight from that page.

If you like that functionality and you want an even more streamlined process there is one more thing we can do.

One Page Checkout

We can make the landing page a little more streamlined by using an extension from WooCommerce.

One Page Checkout (`https://woocommerce.com/document/woocommerce-one-page-checkout/`) will literally add a checkout to any page you want. Users can add items to the cart and checkout all on the same page.

We don't need this extension for our store so we won't be installing it. But it's always good to know you can grow your store in a certain direction if you want or need it.

Now that we have a landing page, and users can add a product to their cart, we can share it with the world and get sales. That's fantastic news!

Building a landing page is great, but we can take it a step further by experimenting with the page. In the web development world, this is usually called **A/B testing**, since we're testing whether version A is better or worse than version B.

Measure and test everything

Now that we have everything set up, it's worth measuring and testing if it works. Oftentimes, when you launch a new product, you don't know what resonates the most with your audience. It's common to have a few guesses and you want to test them.

To collect the best data you can set up an A/B test, where certain users see one version of the landing page and other users see the second version. We can then compare which group had a higher conversion rate, and figure out which marketing message resonates with our audience. However, not everyone has enough traffic to realistically A/B test their pages or their site.

> **How much traffic do you need to A/B test?**
>
> If you want to run actual A/B tests you need quite a bit of data to prove that one variation is better than the other. In e-commerce, the two most common conversion events are add-to-carts and checkouts. You'll need thousands of these events to prove one way or the other which page works better.
>
> If you don't yet have thousands of checkouts, that's okay. Later in this section, we'll recommend other ways to measure and monitor your landing page.

An overview of a CRO experiment

Not every site has enough traffic to set up a conversion rate optimization experiment. Let me show you the one I recently set up. I'm working with a large e-commerce retailer and we're trying to improve the checkout process. One aspect we wanted to improve was the inline error validation.

WooCommerce comes with inline error validation right out of the box. If you're in the checkout and you start filling out fields, you'll probably notice a subtle green and red line on the left side of the input fields. This is a hint that you filled out the field incorrectly.

We wanted to take that subtle hint and make it more obvious. Here's a real A/B test we set up. We're using A/B testing software called **GrowthBook** (`https://www.growthbook.io/`) that lets us set up numerous concurrent experiments.

Figure 12.14 shows what an A/B test looks like in GrowthBook.

Figure 12.14: An A/B test mockup and hypothesis

And, here, as shown in *Figure 12.15*, are the preliminary results.

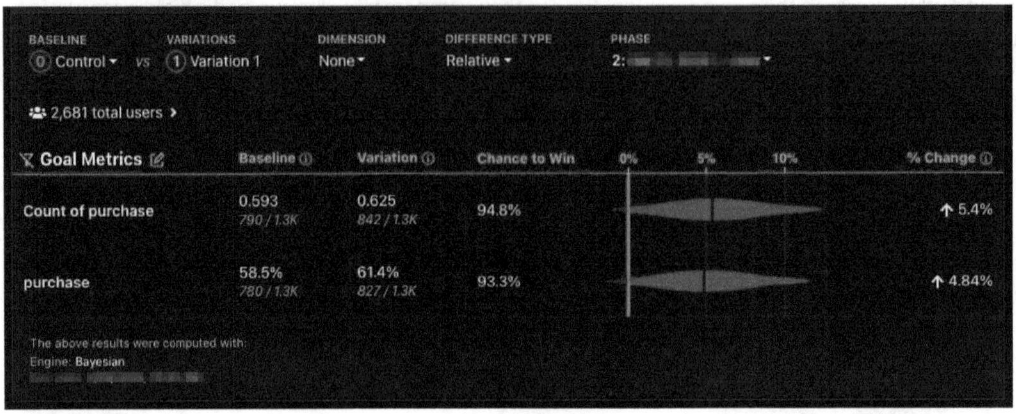

Figure 12.15: A/B test results

You can see that we've had 2,500+ users go through the checkout and, so far, it looks like the new version is a helpful improvement.

About ~62% of users get through the checkout with icons while only ~59% of users get through the checkout without those icons.

So far it looks like this test is successful. However, we're mindful that it's only ~50 more checkouts for the variation. It's still very possible that through random chance, we shuffled more likely buyers into the variation group and the A/B test has little significance. So, while the numbers look good, we're going to keep the test going until we get about 10,000 users.

This is a large e-commerce store and this will take us months. We're lucky we have this much traffic, as many new WooCommerce stores don't have this much traffic so can't run A/B tests like this. For those stores, here's what you can do instead.

Setting up scroll maps and heatmaps

I'm a huge fan of setting up heatmaps and screen recordings. You don't need thousands of visitors. Even if you only had 50 visitors, you can still learn from them.

Before I show you how to set up a heatmap, let me first show you the valuable data you can learn from such a map.

For the *Spring Sale* landing page, we looked at a scroll map and learned that very few users scrolled all the way down the bottom of the products. In fact, only *1/5* of users saw the final product whereas *2/3* saw the first product. See *Figure 12.16*.

Figure 12.16: A scroll map

Our biggest takeaway is that even when you only launch 6 products, you really want to put the most compelling product at the top of the list. Otherwise, users might not see it.

Let's install some software that will help us create scroll maps, heat maps, and other ways to measure the performance of a page.

Follow these steps:

1. In your WordPress admin under **Plugins** | **Add New**, search for hotjar.

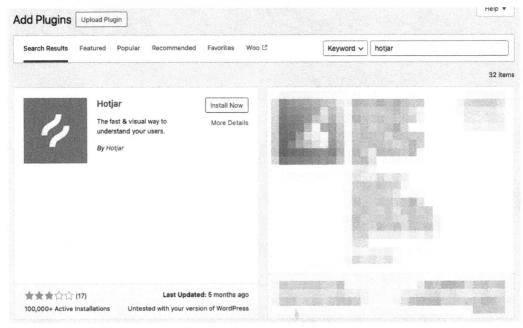

Figure 12.17: Installing Hotjar

2. Activate the plugin.

3. Now let's connect Hotjar to our site. Go to **Settings** | **Hotjar** and we'll see a field called **Hotjar ID** that we can fill out.

4. Sign up for a free account on Hotjar. Then proceed to your list of sites (https://insights. hotjar.com/site/list).

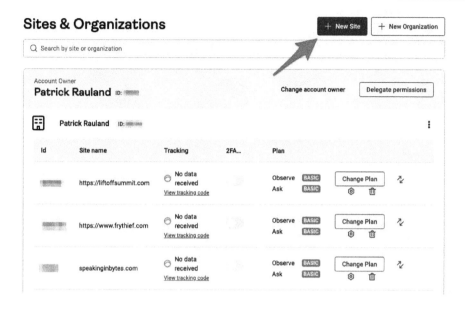

Figure 12.18: Add your site to Hotjar

5. Click **New Site**.

6. Fill out the site information.

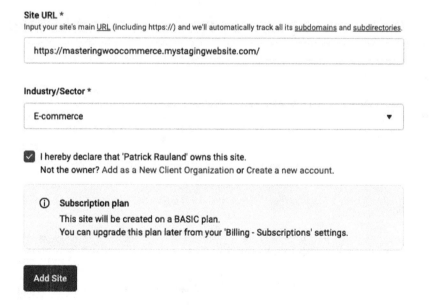

Figure 12.19: Add information for the new site

7.   Now we can view our site in Hotjar and there's an ID we can use.

8.   On the list of sites, you'll see an ID for your site. Copy this.

Id	Site name	Tracking	2FA...	Plan			
	https://liftoffsummit.com	No data received View tracking code		Observe Ask	BASIC BASIC	Change Plan	
	https://www.frythief.com	No data received View tracking code		Observe Ask	BASIC BASIC	Change Plan	
3908512	Mastering WooCommerce	⚠ Not installed Install tracking code		Observe Ask	BASIC BASIC	Change Plan	
	speakinginbytes.com	No data received View tracking code		Observe Ask	BASIC BASIC	Change Plan	

Figure 12.20: Copy site ID from Hotjar

9.   Paste this into your WordPress settings and click **Save**.

10.   You'll then be prompted to verify the installation, which is recommended.

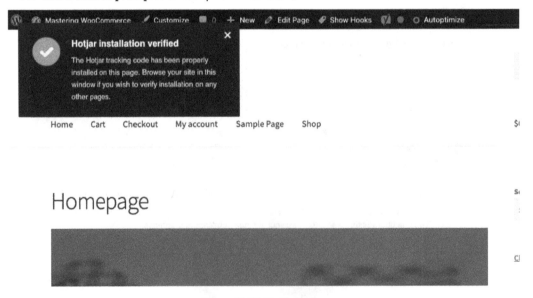

Figure 12.21: Verify Hotjar

And with that, you've installed Hotjar on our site and verified it's working!

You can now set up heatmaps, user recordings, and even more with a paid plan. If this is your first time using technology like this, I recommend setting up all of these. You'll probably find one type of report is much more useful than the others.

I like watching users navigate my site, but it is very time-consuming. Make sure to set up several ways to observe users on your site and you'll quickly figure out what works for you.

Summary

Landing pages are an incredibly important tool for marketing campaigns. Now that you've finished this chapter, you should know how you can build landing pages, how to add e-commerce functionality to landing pages, and how to monitor and experiment with landing pages in order to determine the layout that works best for your audience.

Now that you know how to do this, you can promote and effectively sell your products.

In the next chapter, we'll look into creating and customizing WooCommerce plugins.

13

Creating Plugins for WooCommerce

Ultimately, WooCommerce is a flexible platform not because of the thousands of existing plugins but because you can code your own plugins. You can customize every single line of WordPress and WooCommerce, which means you can literally change anything. Nothing is impossible. It's just a question of how much time you want to invest in coding a solution. It's incredibly powerful for a business.

Becoming a WooCommerce developer could be its own series of books, but we will start by looking at how you can customize a WooCommerce plugin.

We're going to build a basic WooCommerce plugin, look into building integration with WooCommerce, and then customize the edit product page.

The following topics will be covered in this chapter:

- Building a basic WooCommerce plugin
- Customizing order statuses
- Building a settings page with WooCommerce

By the end of this chapter, you should know the basics of building a plugin to customize WooCommerce. First, we need to build a basic WooCommerce plugin.

Technical requirements

This chapter will require you to write code. You should be familiar with PHP code and object-oriented programming to get the most out of this chapter.

The code files for this chapter can be found in the following GitHub repository: `https://github.com/PacktPublishing/Mastering-WooCommerce-/tree/main/Chapter13`.

Building a basic WooCommerce plugin

To get started, we want to build a plugin that will run when WooCommerce is activated on a site. If WooCommerce is not active, we don't want our plugin to run, because that's a waste of processing power.

We're going to create a plugin and then we'll configure it to only load when WooCommerce is active. We need to start by creating a plugin file.

Creating a plugin

To get started, we need to create the plugin files. Plugins are typically stored in the WordPress filesystem under `wp-content/plugins/{your-plugin}`.

There are two ways of creating plugin files:

- Use a single plugin file that contains the entire plugin
- Use a folder with multiple plugin files

The vast majority of plugins use multiple files, so we're going to create a folder for our plugin. Follow the steps given here:

1. Under `/wp-content/plugins/`, add a folder for your plugin. I'm going to call mine `woocommerce-example-plugin`:

Figure 13.1: Our new example plugin folder

2. Navigate inside that folder and create a file with the same name. I'll call mine `woocommerce-example-plugin.php`. This will be the main file for the plugin.

3. Open up your main plugin file in a code editor such as Sublime Text, Visual Studio Code, or Notepad ++.

4. Add the following plugin header information: `https://developer.wordpress.org/plugins/plugin-basics/header- requirements/`

   This information is displayed in your WordPress backend:

```php
<?php
/*
 * Plugin Name: WooCommerce Example Plugin
 * Plugin URI: http://speakinginbytes.com
 * Description: Our custom WooCommerce functionality
 * Version: 2.0
 * Author: Patrick Rauland
 * Author URI: http://speakinginbytes.com
 * License: GPL2
 * License URI: https://www.gnu.org/licenses/gpl-2.0.html
 * Text Domain: woocommerce-example-plugin
 * Domain Path: /languages
 */
```

5. Save the file.

If we stopped right now, we'd have a completely valid plugin that would show up in the backend of WordPress, but it wouldn't do anything. Let's program our plugin so it only runs when WooCommerce is active.

Checking whether WooCommerce is active

We're going to build our plugin from the inside out. We're going to build a class that could do anything and then we're going to build conditions such as checking to make sure WooCommerce is active around the core functionality so it only runs when those conditions are true. Follow these steps:

1. Let's create an empty plugin class. Add the following just beneath the plugin header:

```php
class WC_Example {
    public function __construct(){
        // add more code here
    }
}
// Start running our plugin
$GLOBALS['wc_example'] = new WC_Example();
```

> **Best practice: allow other plugins to refer to your instance**
> The line of code with `$GLOBALS['wc_example'] = new WC_Example();` is useful when you want to allow another plugin to refer to your plugin, but it isn't required.

2. Let's check to make sure WooCommerce is active. In the first version of this book, we had to check the list of active plugins. We wrote code similar to this:

```
// Check to make sure WooCommerce is active
if ( in_array( 'woocommerce/woocommerce.php', apply_
filters('active_plugins', get_option('active_plugins')))))
{

    ((our existing code))

}
```

3. However, WordPress has evolved (https://make.wordpress.org/core/2024/03/05/introducing-plugin-dependencies-in-wordpress-6-5/) and now we don't have to manually check an array of active plugins. Instead, we can now require another plugin to be installed and active with a single line. Add the following, just beneath `Plugin Name: WooCommerce Example Plugin`, to the plugin header:

```
* Requires Plugins: woocommerce
```

By the end, the code should look like this:

```php
<?php
/*
 * Plugin Name: WooCommerce Example Plugin
 * Requires Plugins: woocommerce
 * Plugin URI: http://speakinginbytes.com
 * Description: Our custom WooCommerce functionality
 * Version: 2.0
 * Author: Patrick Rauland
 * Author URI: http://speakinginbytes.com
 * License: GPL2
 * License URI: https://www.gnu.org/licenses/gpl-2.0.html
 * Text Domain: woocommerce-example-plugin
 * Domain Path: /languages
 */
// only run if there's no other class with this name
if ( ! class_exists('WC_Example')){
class WC_Example{
public function __construct(){
// add more code here
```

```
    }
    }
    // Start running our plugin
    $GLOBALS['wc_example'] = new WC_Example();
    }
```

If you look at your list of plugins on your site, you should see something similar to this:

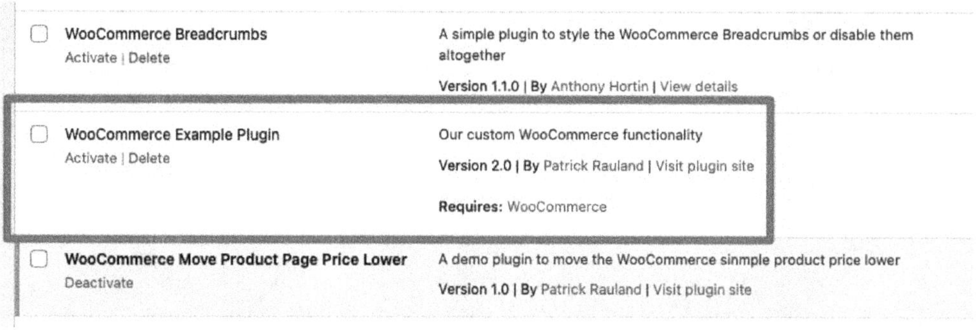

WooCommerce Breadcrumbs	A simple plugin to style the WooCommerce Breadcrumbs or disable them altogether
Activate	Delete	
	Version 1.1.0	By Anthony Hortin	View details
WooCommerce Example Plugin	Our custom WooCommerce functionality
Activate	Delete	Version 2.0	By Patrick Rauland	Visit plugin site
	Requires: WooCommerce
WooCommerce Move Product Page Price Lower	A demo plugin to move the WooCommerce sinmple product price lower
Deactivate	Version 1.0	By Patrick Rauland	Visit plugin site

Figure 13.2: Our plugin requires WooCommerce

Any code you add to the constructor function will only run when WooCommerce is active. This is one of the best ways to start writing your own WooCommerce plugins since it prevents a lot of potential problems.

Now that we know how to write a plugin that works well with WooCommerce and prevents other problems, we can write specific functionality to customize our store.

Customizing order statuses

WooCommerce uses order statuses to track the state of orders. WooCommerce includes several different order statuses, including the following:

- `Completed`
- `Processing`
- `Pending payment`
- `On hold`
- `Refunded`
- `Canceled`
- `Failed`

But some shops may need more order statuses. You might need a *Building* status to indicate that you started building the order. You can easily add a custom order status yourself with some custom code. We're going to start with our WooCommerce example plugin and then we're going to register a post status and add it to WooCommerce.

Using the WooCommerce example plugin

We can start by using the example plugin we created earlier in this chapter.

We'll copy the example code, and then we should change the name of the class from `WC_Example` to `WC_Building_Order_Status`, since this describes what this plugin will do. Here's the code:

```
// only run if there's no other class with this name
if ( ! class_exists('WC_Building_Order_Status')){
    class WC_Building_Order_Status{
        public function __construct(){
            // add more code here
        }
    }
    // Start running our plugin
    $GLOBALS[' wc_building_order_status '] = new WC_Building_Order_
Status();
}
```

> **Note**
>
> We can also add the following code to a class to make it a global variable. This will help other plugins interact with ours. It is purely optional, so you could skip this, but it is a good practice to do so:
>
> ```
> $GLOBALS['wc_building_order_status'] = new WC_Building_Order_
> Status();
> ```

Now we can call two functions from that constructor. We're going to have to add a new post status and then tell WooCommerce where that new post status should show up. We're going to write code to call specific functions that we'll write later.

Add the following to your constructor:

```
// register the new "building" post status
wce_register_building_order_status();
// add "building" to the list of WooCommerce order statuses add_
filter( 'wc_order_statuses', 'wce_add_building_to_order_statuses' );
```

> **Write unique function names**
>
> When you develop a WordPress site, you should code defensively. You never know what another store owner or your future self will install. Perhaps another plugin will use a function called `register_biulding_order_status`. If you had identical function names, your site could crash.
>
> To be a little more defensive, you can prefix your function names. That's what we're doing with our code. I prefixed our function names with `wce`, which stands for "WooCommerce example plugin."

Now that we've called two functions, we need to write both of them.

Registering a post status and adding it to WooCommerce

The first function we called should register a post status. Remember, the post status is what tracks the status of an order. So, we need to add a new post status so we can select it from a dropdown.

This is pretty easy to do if you read through the documentation on the WordPress website about post statuses, available at `https://developer.wordpress.org/reference/functions/register_post_status/`. It only takes a few lines, as you can see here:

```
/**
 * Register the "Building" post status for WooCommerce orders
 */
function wce_register_building_order_status() {
register_post_status( 'wc-building', array(
'label' => 'Building',
'public' => true,
'exclude_from_search' => true,
'show_in_admin_all_list' => true,
'show_in_admin_status_list' => true,
'label_count' => _n_noop( 'Building <span class="count">(%s)</span>',
'Building <span class="count">(%s)</span>' ) )
);
}
```

With this function, we created a new post status called `Building` and we also hid it from searches and from a few other areas in your WordPress site.

Now we need to tell WooCommerce when to display this new order status. We'll loop through each of the existing order statuses in WooCommerce and add our new order status right after wc-processing, which is the Processing status:

```
/**
 * Add the new "Building" post status to the list of WooCommerce order
statuses
 */
function wce_add_building_to_order_statuses( $order_statuses ) {
$new_order_statuses = array();
// add new order status after processing
foreach ( $order_statuses as $key => $status ) {
$new_order_statuses[ $key ] = $status;
if ( 'wc-processing' === $key ) {                $new_order_statuses['wc-
building'] = 'Building';
}
}
return $new_order_statuses;
}
```

With this code, we're looping through all of the order statuses. If we find one called wc- processing, we add a new order status called **Building**.

When you're done, save your files, make sure your plugin is activated in the backend, and then open up an order. You should see your new order status, as we can see in the following screenshot:

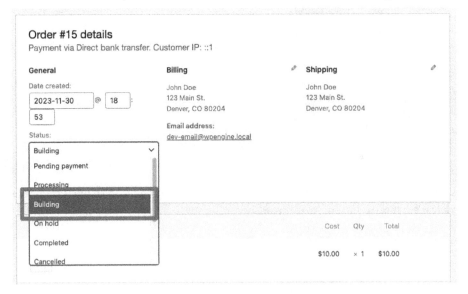

Figure 13.3: A new custom order status

Now we can change any order to the **Building** status. This will help us track the status of orders. In this example, we added one order status, but you could add as many as you want to help you track your orders.

Next up, we will look at building integration with WooCommerce.

Building a settings page with WooCommerce

When you're building your own plugin, you'll very likely have to build a settings page. There's a lot to building a settings page from scratch. But if you are just building an integration, WooCommerce has created some technology that makes it easier for you to add your own settings screen.

We're going to cover some of the coding decisions in the WooCommerce `Integration` class. If you want, you can see the finished code at `https://developer.woocommerce.com/docs/creating-custom-settings-for-woocommerce-extensions/`.

We can see that one of the plugins we looked at earlier in this book takes advantage of the `Integration` class, which we can see in the following screenshot:

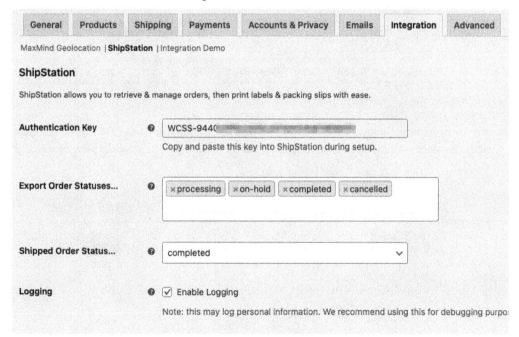

Figure 13.4: ShipStation uses the WooCommerce Integration class

We're not going to explain every single line. But we are going to cover the essential lines and why you need them.

Let's build our own settings page using the `Integration` class. We'll start by creating the main file.

Creating the main integration file

The main integration file will activate all other functionality in the `Integration` class. So, let's start with that. Follow the steps given here:

1. Let's create a new plugin folder in `/wp-content/plugins/`. We can call it `woocommerce-integration-demo`.

2. Create `woocommerce-integration-demo.php`.

3. Open that file.

4. Paste in the following plugin header:

```
/**
 * Plugin Name: WooCommerce Integration Demo
 * Requires Plugins: woocommerce
 * Plugin URI: https://github.com/BFTrick/woocommerce-
integration-demo
 * Description: A plugin demonstrating how to add a new
WooCommerce integration.
 * Author: Patrick Rauland
 * Author URI: http://speakinginbytes.com/
 * Version: 2.0
 */
```

5. Now we have the real work to do. Let's first check whether the WooCommerce `Integration` class exists. If it does, we can include it and then call our own function to register our own integration:

```
if ( class_exists( 'WC_Integration' ) ) {
    // Include our integration class.
    include_once 'includes/class-wc-integration-demo-
integration.php';
    // Register the integration.
    add_filter( 'woocommerce_integrations', array( $this, 'add_
integration' ) );
}
else {
    // throw an admin error if you like
}
```

We called a function (`add_integration`) to register our own function. We need to tell WooCommerce what to do with this file. We can add our integration to the list of integrations that WooCommerce automatically loads. Let's add the name of our integration to the array of integrations:

```
/**
 * Add a new integration to WooCommerce.
 */
public function add_integration( $integrations ) {
  $integrations[] = 'WC_Integration_Demo_Integration';
  return $integrations;
}
```

That's pretty much it for this first file. We loaded our own integration file and told WooCommerce to add it to the list of integrations.

When we're done, it should look like this:

```
<?php
/**
 * Plugin Name: WooCommerce Integration Demo
 * Requires Plugins: woocommerce
 * Plugin URI: https://github.com/BFTrick/woocommerce-
integration-demo
 * Description: A plugin demonstrating how to add a new
WooCommerce integration.
 * Author: Patrick Rauland
 * Author URI: http://speakinginbytes.com/
 * Version: 2.0
 */
if ( ! class_exists( 'WC_Integration_Demo' ) ) :
class WC_Integration_Demo {
  /**
   * Construct the plugin.
   */
  public function __construct() {
    add_action( 'plugins_loaded', array( $this, 'init' ) );
  }

  /**
   * Initialize the plugin.
   */
  public function init() {
    // Checks if the WooCommerce Integration class exists
    if ( class_exists( 'WC_Integration' ) ) {
      // Include our integration class.
```

```
        include_once 'includes/class-wc-integration-demo-
integration.php';
        // Register the integration.
        add_filter( 'woocommerce_integrations', array( $this,
'add_integration' ) );
      }
      else {
        // throw an admin error if you like
      }
    }


    /**
     * Add a new integration to WooCommerce.
     */
    public function add_integration( $integrations ) {
      $integrations[] = 'WC_Integration_Demo_Integration';
      return $integrations;
    }
  }
  $WC_Integration_Demo = new WC_Integration_Demo( __FILE__ );
  endif;
```

The preceding code creates the basics of our integration and tells WooCommerce to load our integration. If we look at our list of plugins, we'll see a new plugin.

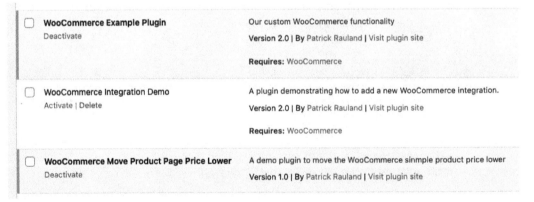

Figure 13.5: Our plugin appears in the plugin list

Note

Don't activate the plugin quite yet. There will be some PHP errors since we haven't written all of the code. If you ever activate a plugin with an error, you can use an FTP plugin to delete the plugin files to prevent those errors, log back into WordPress, and disable the plugin.

Now we need to actually create our own integration file, which will specify exactly what settings we need.

Creating the Integration child class

The way we're creating our settings page is by taking advantage of a concept in programming called **inheritance**. Inheritance means we're using a child class that inherits programming from the parent class. Basically, WooCommerce wrote the parent `Integration` class and we're going to write a child class that will automatically have certain things defined, which will save us dozens of hours of coding.

There are two important pieces of the `Integration` child class. The first part is the constructor and the second is where we define our settings. Let's look at the constructor first.

Creating a constructor

The constructor is where we declare all of our integration information, load our settings, and add our own variables that the user can interact with. Follow these steps:

1. Let's start by creating the `class-wc-integration-demo-integration.php` file, which we can add to a new `includes` folder within `woocommerce-integration-demo`.

   Now we can start writing code. Let's add an empty constructor:

   ```
   /**
    * Init and hook in the integration.
    */
   public function __construct() {
   global $woocommerce;
   }
   ```

2. Now let's configure some integration settings. Add the following code to your constructor right after `global $woocommerce`:

   ```
   $this->id = 'integration-demo';
   $this->method_title = __( 'Integration Demo', 'woocommerce-
   integration-demo' );
   $this->method_description = __( 'An integration demo to show
   you how easy it is to extend WooCommerce.', 'woocommerce-
   integration-demo' );
   ```

3. The rest of the constructor is for our settings. We need to load the settings, add fields, process our options, and sanitize the options. Add the following after the previous code:

   ```
   // Load the settings.
   $this->init_form_fields();
   $this->init_settings();

   // Define user set variables.
   ```

```php
$this->api_key = $this->get_option( 'api_key' );
$this->debug = $this->get_option( 'debug' );

// Actions.
add_action( 'woocommerce_update_options_integration_' . $this-
>id, array( $this, 'process_admin_options' ) );

// Filters.
add_filter( 'woocommerce_settings_api_sanitized_fields_' .
$this->id, array( $this, 'sanitize_settings' ) );
```

4. We've finished everything we need for the constructor. Let's put the whole block of code together and then we'll dig into what we accomplished in the following section.

```php
/**
 * Init and hook in the integration.
 */
public function __construct() {
    global $woocommerce;

    $this->id                 = 'integration-demo';
    $this->method_title       = __( 'Integration Demo',
'woocommerce-integration-demo' );
    $this->method_description = __( 'An integration demo to
show you how easy it is to extend WooCommerce.', 'woocommerce-
integration-demo' );

    // Load the settings.
    $this->init_form_fields();
    $this->init_settings();

    // Define user set variables.
    $this->api_key = $this->get_option( 'api_key' );
    $this->debug   = $this->get_option( 'debug' );

    // Actions.
    add_action( 'woocommerce_update_options_integration_'
. $this->id, array( $this, 'process_admin_options' ) );

    // Filters.
    add_filter( 'woocommerce_settings_api_sanitized_fields_' .
$this->id, array( $this, 'sanitize_settings' ) );
}
```

The preceding code tells WooCommerce the `title`, `description`, and `settings` fields we want to see on the page. Now that we have told WooCommerce that we want settings, we have to define which settings. Let's write our field settings.

Adding field settings

Now that we have our basic integration setup, we need to add our fields. So far, everything we've written will work for just about any integration. Now we're going to write code for the specific settings our integration needs. Follow these steps:

1. Let's start with a function to create the fields:

    ```
    /**
     * Initialize integration settings form fields.
     *
     * @return void
     */
    public function init_form_fields() {
        // we'll add more code here
    }
    ```

2. Now we can add the settings to this function. Let's add one setting first. Add a field for our API key. We're going to create an array called `form_fields` and add all of our fields to that array:

    ```
    $this->form_fields = array(
        'api_key' => array(
            'title'       => __( 'API Key', 'woocommerce-
    integration-demo' ),
            'type'        => 'text',
            'description' => __( 'Enter with your API Key. You can
    find this in "User Profile" drop-down (top right corner) > API
    Keys.', 'woocommerce-integration-demo' ),
            'desc_tip'    => true,
            'default'     => '',
        ),
    );
    ```

The preceding code is for an API key. The three important fields are as follows:

*   `title`: The name of the field
*   `type`: This is a text field (as opposed to a paragraph or number input)
*   `description`: This shows up next to the text field to explain what it is to the user

1. Now that we know how to add one field, you can add the rest of the fields. WooCommerce has a whole section in the documentation about these fields. If you want to know the types of fields you can display and what each setting means, read through the documentation at `https://developer.woocommerce.com/docs/creating-custom-settings-for-woocommerce-extensions/#1-creating-settings` s.

The following code shows the complete array when you finish adding fields:

```
$this->form_fields = array(
    'api_key' => array(
        'title'       => __( 'API Key', 'woocommerce-
integration-demo' ),
        'type'        => 'text',
        'description' => __( 'Enter with your API Key. You can
find this in "User Profile" drop-down (top right corner) > API
Keys.', 'woocommerce-integration-demo' ),
        'desc_tip'    => true,
        'default'     => '',
    ),
    'debug' => array(
        'title'       => __( 'Debug Log', 'woocommerce-
integration-demo' ),
        'type'        => 'checkbox',
        'label'       => __( 'Enable logging', 'woocommerce-
integration-demo' ),
        'default'     => 'no',
        'description' => __( 'Log events such as API requests',
'woocommerce-integration-demo' ),
    ),
    'customize_button' => array(
        'title' => __( 'Customize!', 'woocommerce-integration-
demo' ),
        'type' => 'button',
        'custom_attributes' => array(
            'onclick' => "location.href='http://www.woothemes.com'",
        ),
        'description' => __( 'Customize your settings by going to
the integration site directly.', 'woocommerce-integration-demo'
),
        'desc_tip' => true,
    ),
);
```

This code configures the settings for four settings. This is similar to the previous block of code where we configured one settings field. I wanted to show you what it looks like when you have multiple fields. Notice the different field types, such as `button` and `checkbox`.

All of the settings fields are added to the `init_form_fields` method.

2. When you're all done, and you load the **Integration** page, it will look something like the following screenshot:

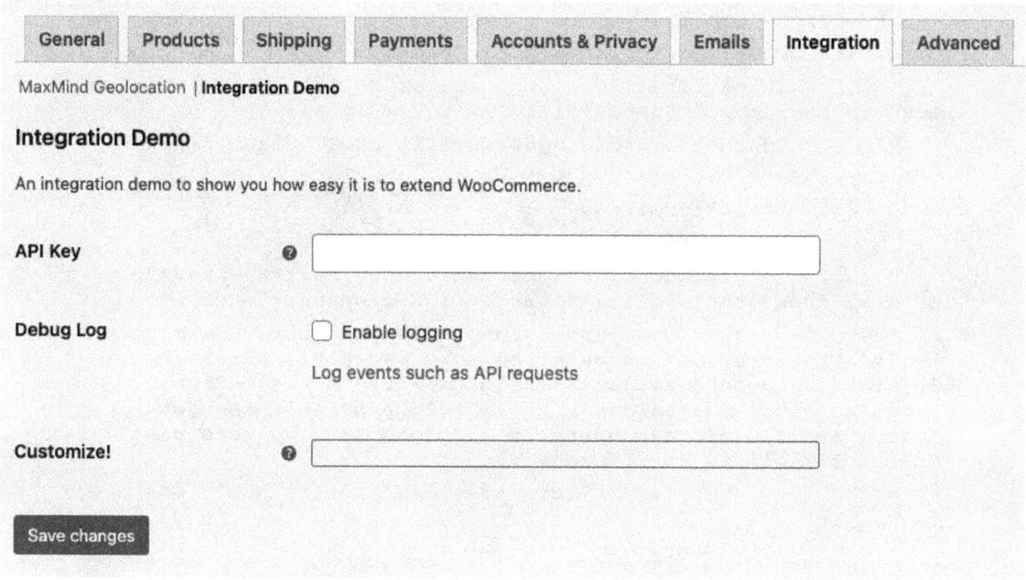

| General | Products | Shipping | Payments | Accounts & Privacy | Emails | Integration | Advanced |

MaxMind Geolocation | **Integration Demo**

Integration Demo

An integration demo to show you how easy it is to extend WooCommerce.

API Key   ❷   [   ]

Debug Log   ☐ Enable logging

                            Log events such as API requests

Customize!   ❷   [   ]

Save changes

Figure 13.6: Our settings page is starting to come together

This is close to what we want. We can see two of our settings pretty clearly. But the **Customize!** button setting we added doesn't look right. That's because WooCommerce doesn't include code to display a button. But we can add our own display code.

3. Add the following code beneath the `init_form_fields` function:

```
/**
 * Generate Button HTML.
 */
public function generate_button_html1( $key, $data ) {
  $field    = $this->plugin_id . $this->id . '_' . $key;
  $defaults = array(
      'class'              => 'button-secondary',
      'css'                => '',
      'custom_attributes'  => array(),
      'desc_tip'           => false,
      'description'        => '',
      'title'              => '',
  );
```

```php
$data = wp_parse_args( $data, $defaults );

ob_start();
?>
<tr valign="top">
    <th scope="row" class="titledesc">
        <label for="<?php echo esc_attr( $field ); ?>"><?php
echo wp_kses_post( $data['title'] ); ?></label>
        <?php echo $this->get_tooltip_html( $data ); ?>
    </th>
    <td class="forminp">
        <fieldset>
            <legend class="screen-reader-text"><span><?php
echo wp_kses_post( $data['title'] ); ?></span></legend>
            <button class="<?php echo esc_attr( $data['class']
); ?>" type="button" name="<?php echo esc_attr( $field );
?>" id="<?php echo esc_attr( $field ); ?>" style="<?php
echo esc_attr( $data['css'] ); ?>" <?php echo $this->get_
custom_attribute_html( $data ); ?>><?php echo wp_kses_post(
$data['title'] ); ?></button>
            <?php echo $this->get_description_html( $data );
?>
        </fieldset>
    </td>
</tr>
<?php
return ob_get_clean();
}
```

That will make our button look a lot better.

Our code works as long as users input perfect data. But that *never* happens. Users will inevitably enter bad data. We need to sanitize our data.

We already added some code to call a sanitization function to our constructor. That's what this line does:

```php
add_filter( 'woocommerce_settings_api_sanitized_fields_' .
$this->id, array( $this, 'sanitize_settings' ) );
```

4. Let's add our sanitization function. Add the following code beneath the `generate_button_html` function:

```php
/**
 * Santize our settings
 * @see process_admin_options()
```

```
*/
public function sanitize_settings( $settings ) {
// We're just going to make the api key all upper case
characters since that's how our imaginary API works
if ( isset( $settings ) &&
     isset( $settings['api_key'] ) ) {
$settings['api_key'] = strtoupper( $settings['api_key'] );
}
return $settings;
}
```

This is very simple code that changes any lowercase characters in an API key to uppercase. You can take this much further by using WordPress and PHP functions to clean and process your data to make sure it's all saved correctly.

There's a huge list of sanitization functions in the WordPress documentation: `https://developer.wordpress.org/apis/security/sanitizing/`

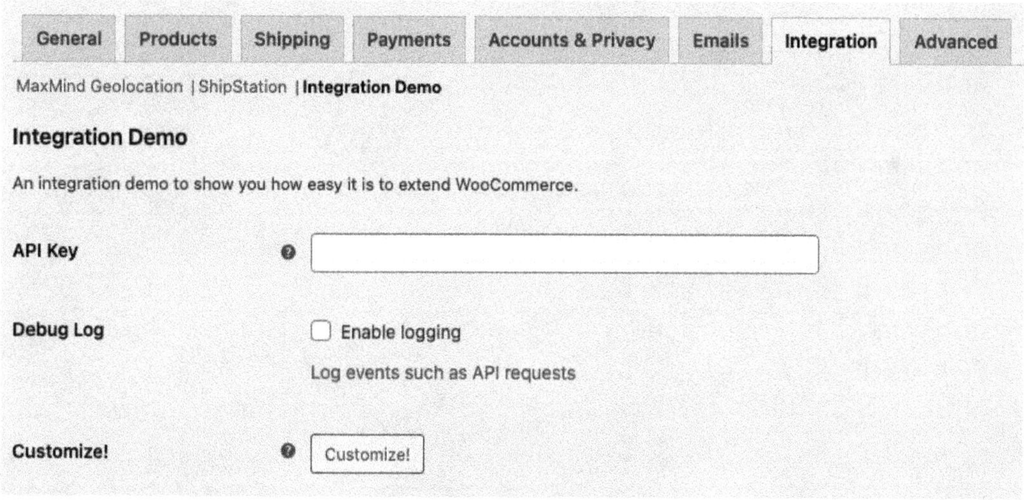

Figure 13.7: Our completed integration settings page

This is some of the simplest code I can show you to customize WooCommerce. But it's still pretty complex for new coders. This code might take a seasoned developer an hour or more to figure out and customize to suit their needs. So don't worry if it takes you a while to understand. I highly recommend that you read through the documentation, available at `https://developer.woocommerce.com/docs/creating-custom-settings-for-woocommerce-extensions/`, so you can fully understand it.

Summary

WooCommerce is an incredibly open system where you can do just about anything. We covered some simple tutorials in this chapter to give you a taste of what you can do. You should now know how to build a plugin that only works when WooCommerce is active, how to add new custom order statuses, and how to build an integration with another service.

You've taken the first steps to mastering WooCommerce and maybe even developing your own themes and plugins for WooCommerce. In the next chapter, we'll take a peek into a few advanced topics to become an informed developer who makes safer, more accessible websites.

14

Next Steps with WooCommerce

We've built and customized every aspect of an e-commerce business built on top of WooCommerce. In this chapter, I want to provide you with three ways you can continue to develop your skills. All of these are important topics in very different ways and could be how you specialize as an e-commerce developer.

Security is important for any business; building an inaccessible website is risky and isn't the right choice, and if you are building your business on top of WooCommerce, you want to have some agency over the direction it goes in.

We're going to cover the following topics:

- Why and how to make your WooCommerce Store accessible
- Keeping WooCommerce safe and secure
- Staying up to date with WooCommerce and open source software

Once you've explored these topics, you'll know more about WooCommerce than most developers.

Technical requirements

This chapter is a bit more forward-looking. We're going to cover the following topics at a slightly higher level.

Every one of these topics will be important for a WooCommerce store owner.

If you're a developer who understands WordPress code, there will be even more you can do in this chapter.

We'll be digging into GitHub (`https://github.com/PacktPublishing/Mastering-WooCommerce-/tree/main/Chapter14`) to manage a code base. It is not required, but it may be helpful to create a GitHub account.

Why and how to make your WooCommerce store accessible

Web accessibility is the practice of designing and developing websites in a way that ensures that people with disabilities can perceive, understand, navigate, and interact with the content effectively, regardless of their abilities or impairments.

In the e-commerce world, there are four major reasons to care about accessibility:

- Inclusivity
- Legal liability
- Business benefits
- Preparing for demographic trends

Inclusivity

One criticism of accessibility that you'll see online is this: why spend so much effort to make a website accessible to blind users? Couldn't you spend that same effort to grow your business even more?

Everyone will have a disability at some point

If you look narrowly at the term *disabled*, I can understand the roots of this argument. But if you look a little more long-term, everyone will be disabled at some point or in some contexts:

- **Everyone ages**: Eyes age, small words become harder to read, and contrast becomes more important. If you're so fortunate as to live long enough, you may have arthritis and find it difficult to move a mouse.
- **Bones break**: Have you ever had your arm in a cast and tried to navigate a website without a mouse? It's tricky unless you spend some time making it keyboard accessible.
- **Slow connection**: We've all had to access important information on a mobile phone but the internet connection was miserably slow. It would be nice to have alt text to describe an image while it's loading.
- **Environmental factors**: If you're in a loud bar and you need to hear sound in a video, what do you do? Captions or a transcript would be perfect so you don't have to leave the room. If you're standing outside in the bright sun, how easy is it to read your website?

When you think about accessibility through that lens, you aren't designing a website for other people. You're building the web in the way your future self would be proud of.

Around ~11% of the population in the United States have a visual or hearing disability (source: `https://www.cdc.gov/ncbddd/disabilityandhealth/infographic-disability-impacts-all.html`). This group represents a market segment with disposable income and one that wants to get things done online from the comfort of home.

By building an accessible website, you're helping as many people as you can and unlocking your website for a large group of potential customers.

Legal liability

It's common knowledge in the e-commerce world that litigation around accessibility is increasing. There are more and more lawsuits for inaccessible websites.

UsableNet has a *2023 Year End Report* (`https://info.usablenet.com/2023-year-end-digital-accessibility-lawsuit-report-download-page`) full of data. Here are a few data points:

- Over the last five years, 82% of the top 500 e-commerce websites received an ADA-based lawsuit (`https://www.3playmedia.com/blog/key-takeaways-usablenets-ada-web-app-report/`).

- Most of the more than 4,000 ADA web accessibility lawsuits filed in 2022 and the nearly 16,000 ADA web accessibility lawsuits filed in the last five years had plaintiffs with visual disabilities. Plaintiffs with visual disabilities filed the most cases, and plaintiffs with auditory disabilities came in second; most often, claims filed by people with hearing disabilities focused on digital accessibility issues for video, such as a lack of captions or missing audio descriptions. Source: `https://blog.usablenet.com/7-facts-your-ecommerce-company-should-know-about-ada-web-lawsuits`.

The minimum standards for websites have gradually and then suddenly risen. When I started over 10 years ago, people were just surprised that we could build a website that could accept real money. E-commerce is now part of everyday life, and as a society, we have high expectations when it comes to helping the disabled.

This is also the part where I put on my developer hat. I am not a legal expert and my advice should not be considered legal advice. However, I have worked with accessibility experts and I have reduced legal liability in the past. So I'll share the 80/20 of what I've learned.

Don't attract the attention of the big firms

There are 10 law firms that are responsible for over 80% of all ADA cases (source: `https://www.forbes.com/sites/gusalexiou/2024/01/09/new-york-led-the-way-in-us-web-accessibility-lawsuits-in-2023-report-shows/`).

These firms know which types of lawsuits will win and pay out. It's a business for them. If you can make your website more accessible than your competitors', you'll likely be ignored by the big firms. They'd prefer to spend their time suing a company with a stronger case.

Business benefits

If you're working on making your website more accessible, you'll notice you'll also improve other elements, such as **search engine optimization (SEO)** and **conversion rate optimization (CRO)**.

Improving SEO

By adding alt text to images, improving heading structure, and uncovering problems in the HTML markup, we can make some gains in SEO. Let's dig into some of these strategies:

- **Adding alt text to images**: It's easy to skip over adding alt text to images. People want to move fast and it can feel like you're adding data that no one will ever see. But as we mentioned before, many people in the US have a visual impairment or slow internet, and it would be useful for those people to have alt text load for images. (Source: `https://moz.com/learn/seo/alt-text`.)

- **Improving heading structure**: Headings add a hierarchical structure to your page. Disabled users are used to navigating by pulling up a list of all headings and navigating to the heading that's closest to what they want. I've uncovered hidden H1 headings in the source code of some older themes. By clearing this up for accessibility, we're also improving our page structure for Google. (Source: `https://yoast.com/how-to-use-headings-on-your-site/`.)

- **Uncovering problematic HTML**: When you hire developers to help you write custom code or if you install a plugin or theme from an inexperienced author, you may notice problematic HTML. I've noticed H2 headings in a cart at times – and that cart widget appeared on every page of the site. Similarly, I've used an FAQ plugin that was completely inaccessible to keyboard users. It was primarily the HTML that was the culprit. In that specific example, I emailed the developer and they fixed the issue. By clearing up incorrect HTML, you'll greatly help Google understand your site.

Improving CRO

One of the changes I recently made to a WooCommerce site was adding icons to checkout fields to show users' mistakes. By default, in WooCommerce core, they only show you a subtle red line.

After making this change, we reviewed screen recordings and we observed users pre-emptively fixing fields (such as the email field) before pressing submit. This change, which we made for accessibility purposes, protects us legally, and it also should increase conversions over the long term.

As you dig into accessibility, most of these changes will have subtle but compounding effects on your conversion rate.

Prepare for demographic trends

The largest generation of Americans, the Baby Boomers, are in the middle of retiring. There's a huge increase in the number of people who are 60+ years old in the US. To know more about this, you can visit this link: `https://www.census.gov/library/stories/2023/05/2020-census-united-states-older-population-grew.html`.

As the Boomers pass 60 years old, more and more of them will have trouble with their vision and hearing. And even if they're not in your target demographic, they still spend money on their children and grandchildren, so they're still your customers.

If you're thinking long term, the Millenials are the next-largest generation; the oldest of their generation are in their 40s and starting to need reading glasses.

If you don't think you have enough customers to justify needing an accessible website now, you should prepare your business for the future and start making your website accessible now.

How to make your store accessible

In the United States, your store needs to be WCAG 2.1 compliant (source: `https://www.w3.org/TR/WCAG21/`).

In Europe, in 2025, you will have to be compliant (source: `https://userway.org/blog/european-accessibility-act/`).

The Europeans still need to define their upcoming requirements. The WCAG requirements for the US read like a technical manual from the early 2000s. It's hard for a layperson to understand what you need to do to have an accessible store.

Let me share some of the simple things you can do, even if you can't code. These two things will greatly help your WooCommerce store and give you the first layer of protection from legal action.

Improve color contrast

One of the things you control through your theme and the customizer is the colors you use. Think about someone who has low vision, whose reading glasses are across the room, or who is using their cell phone outside in the direct sunlight. Can they read your text? Probably not.

Having strong color contrast will greatly help. And it is part of the WCAG requirements. If you navigate to `https://webaim.org/resources/contrastchecker/`, you can use their free contrast checker.

Enter the text color and the background color and see whether your text is accessible (WCAG AA).

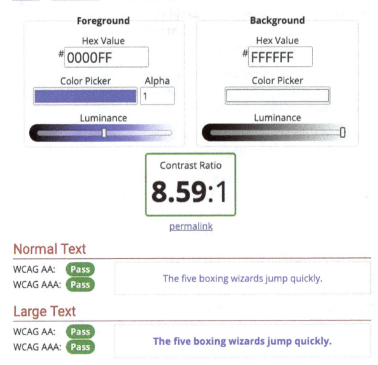

Figure 14.1: A WCAG AAA contrast ratio

I recommend testing the following:

- All text on your home page

- The standard link color text on single blog post pages

- The add-to-cart button on your product pages

- Buttons in the checkout

Technically, all text on your site needs to meet contrast requirements, but these pages are a great place to start.

Test your store using the keyboard

Did you know you can use a keyboard to navigate a website? Give it a try right now: you can use *Tab* and *Shift + Tab* to navigate between interactive elements. You can also press *Enter* to press a button or click a link.

As you navigate through your site, you should see an outline around the element you have selected.

Figure 14.2: An outline surrounds any element with keyboard focus

If you don't see an outline, then you will find it almost impossible to navigate your store by keyboard. So, we need to add an outline when users are navigating by keyboard.

If you don't see a visible outline around elements when you start tabbing through your site, you probably have CSS that's hiding the outline. Style sheet resets were common a decade ago in web development and some of these turned off outlines. There are still projects that might have unintentionally reset their outline.

If you know how to write CSS code, you can add this yourself:

```
:focus-visible {
outline: 2px solid;
outline-offset: 2px;
}
```

Notice we're setting a 2px outline. But we're not specifying the color. That's actually intentional. In most cases, the browser detects the background color and sets the outline color to a contrasting color, effectively deciding what color to show.

Even spending a few hours working on the accessibility of your website and tackling one of these tests will help a very large segment of the population and protect your business.

Now let's talk about how you can keep your store safe and secure in the digital realm.

Keeping WooCommerce safe and secure

This whole book has been guiding you to build an incredible WooCommerce store that will make a lot of money for you or your client. But that store won't do you any good if hackers can get easy access to your site and siphon money from your business or turn the whole website into an ad for cryptocurrency.

There are several steps you can take that make your site much more resilient.

Check your hosting before you launch

Before you launch your store, it's worth double-checking that your website host is both reputable and has modern technology.

We talked about hosts back in *Chapter 9, Speeding Up Your Store*. It's worth talking about website hosts again in the context of security. A **website host** is the infrastructure for your site. If you want to have a secure website with an insecure host, it's sort of like rolling up the windows of your car but leaving the door unlocked.

Here are some of the features I'd recommend you look for in a host:

- Actively monitor malware
- Keep WordPress core updated automatically
- Uses a supported version of PHP *(more on this shortly)*
- Back up your website daily
- Employs a state-of-the-art web application firewall
- Provides free **Secure Sockets Layer (SSL)** certificates

There are many great WordPress hosts. Ultimately, look for one that cares about WordPress and upholding best practices such as keeping PHP and WordPress updated. Ultimately, if a good host costs an extra $10 a month than a cheaper host, it's probably worth it to build this e-commerce business on a solid foundation.

If you want to dig a little more into how a host should handle security, this blog post has some good information: `https://pressable.com/blog/hacking-how-pressable-handles-it/`.

> **Launch your store first, then make it secure**
>
> I purposefully put security near the end of this book. Other than choosing a host, many of the choices you make for securing your website can be done after launch. So, my recommendation is to choose a good host, and then do all of the rest security upgrades after you launch.
>
> Alternatively, you could even wait to see whether your e-commerce idea takes off and then improve your security.

Use an SSL certificate

One essential element for e-commerce security is an SSL certificate. An **SSL certificate** is the technology you need to have your website load in HTTPS. The vast majority of payment gateways will require HTTPS to protect sensitive payment information.

Your host should be able to set up an SSL for you for free.

Keep WordPress core and plugins up to date

WooCommerce handles all of the e-commerce technology and functionality for your site. That's what the WooCommerce developers do all day—make sure that functionality is solid.

Under the hood is WordPress, which handles foundational security. Some hosts will automatically update WordPress for you; if you don't like hosts updating plugins for you, then make sure you build a practice of updating WordPress regularly.

This of course also applies to WooCommerce and all of your WordPress plugins. When you have enough plugins on your website, you'll often see a huge list of updates. Don't worry about updating them every day. Instead, have a cadence that you can keep up with. I'm a big fan of updating WordPress and all of your plugins on a monthly or quarterly cadence.

If you're looking for a robust solution that patches your website immediately when a security fix is available, you can look into services like Patch Stack: `https://patchstack.com/`.

Keep your version of PHP supported

One thing some website hosts haven't been great at is supporting modern versions of PHP. PHP is the programming language behind WordPress and it updates just like WordPress. From a user's perspective, you won't notice the differences, but older versions of PHP are **end-of-life** (**EOL**), meaning they no longer receive security updates.

You don't want to build your business in an environment where they don't fix security holes. Here's a graph showing the supported versions of PHP from their website:

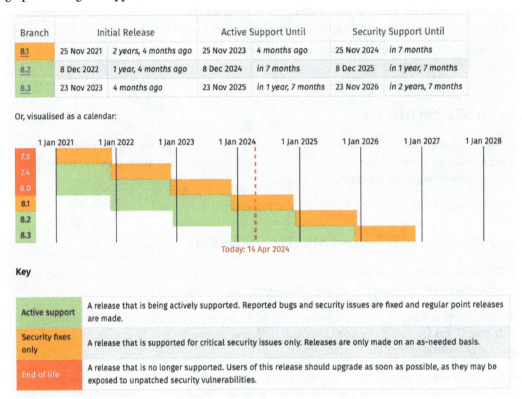

Branch	Initial Release		Active Support Until		Security Support Until	
8.1	25 Nov 2021	2 years, 4 months ago	25 Nov 2023	4 months ago	25 Nov 2024	in 7 months
8.2	8 Dec 2022	1 year, 4 months ago	8 Dec 2024	in 7 months	8 Dec 2025	in 1 year, 7 months
8.3	23 Nov 2023	4 months ago	23 Nov 2025	in 1 year, 7 months	23 Nov 2026	in 2 years, 7 months

Or, visualised as a calendar:

Today: 14 Apr 2024

Key

Active support	A release that is being actively supported. Reported bugs and security issues are fixed and regular point releases are made.
Security fixes only	A release that is supported for critical security issues only. Releases are only made on an as-needed basis.
End of life	A release that is no longer supported. Users of this release should upgrade as soon as possible, as they may be exposed to unpatched security vulnerabilities.

Figure 14.3: A graph showing how quickly PHP versions release and reach EOL

You can see the full list of supported versions of PHP on their website: `https://www.php.net/supported-versions.php`.

If you're building a website today, I'd only build it with PHP 8.2 or greater. The other versions are already too old and won't be getting the attention you need.

Two-factor login for administrators

We've talked a lot about updating your core software. But there's more than updates that you should do.

One of the best security features you can add is **two-factor authentication**. This forces a user to type in a code on their phone in addition to their password. With a two-factor authentication setup, if a hacker gets a password, they still can't log in.

The hardest part of two-factor authentication

Honestly, the hardest part of using two-factor authentication is dealing with the annoyance. You always need your phone near you and with an internet connection. I've set up two-factor authentications on some websites that make millions of dollars a year, and a month after I set up the security, the owner disabled it because they found it annoying.

So, while installing a plugin solves the technical problem, you also need to solve the people problem, which is done by educating them as to why it's great to protect your business.

You can see a huge list of plugins by searching for two-factor on the WordPress.org plugin repository: `https://wordpress.org/plugins/search/2-factor/`.

One plugin I've used in the past is WP 2FA (`https://wordpress.org/plugins/wp-2fa/`).

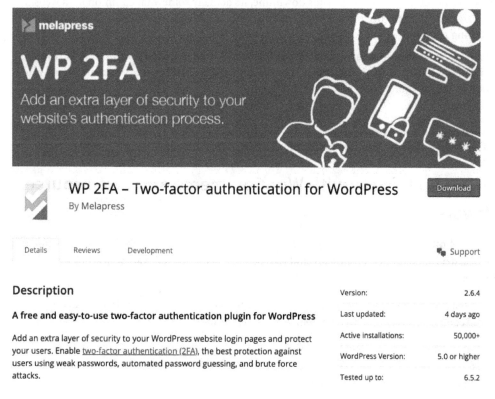

Figure 14.4: The WP 2FA plugin

This gives you the ability to add two-factor authentication for your administrators, and you can also send the code to an email address if you prefer that to your phone.

Scan for downtime

E-commerce websites run 24/7 and should be accepting payment at all hours of the day. Of course, humans can't stay up 24/7. That's why we need technology to let us know when the website is down and whether it's down long enough to be of concern.

There are many services that can monitor the status of your website, but since we've already covered Jetpack in this book, I recommend the downtime monitoring service built into Jetpack (`https://jetpack.com/support/monitor/`).

It's easy and free to set up, and it's a great place to start.

How long can my website be down?

If you've never monitored a website's downtime, it's actually somewhat common for a website to go down for 5 minutes at random hours such as 2 am. Don't worry about the small outages. Nothing is up 100%.

But if your website is regularly going down, then you'll want to talk to your host. Plus, if you're down for a long period of time when your store should be making lots of money, you'll also want to talk to your host.

Now that we know how to keep our store safe, let's talk about how we can stay up to date with WooCommerce and how you can contribute to open source software.

Staying up to date with WooCommerce and open source software

When you build your business on a particular piece of software, it's worth the time investment to follow the development of that software and have an idea where the software is heading over the next 1–3 years.

Besides just staying up to date with software, WooCommerce is open source. That means anyone can view the code and can choose to contribute new and improved code.

Contributing code is something for developers and engineers, but staying up to date with software is useful for anyone running an e-commerce business, even if you aren't a coder. Let's start with the ways you can stay up to date with WooCommerce.

Follow the Developer Blog

WooCommerce actually does a really good job of communicating where the software is going through their Developer Blog: `https://developer.woocommerce.com/blog/`.

Figure 14.5: Woo Developer Blog

This blog is how I find out about upcoming features in WooCommerce; if they need any testing on my end, I'll add those tests to my backlog so I know at some point I need to work on them.

A big focus for the past few years is on WooCommerce blocks. If you followed the blog back then, you'd know not only what has already come out, but also what will be coming out in the future and how those features will work technically. If you know an important feature is coming out, you can save time and money by *not* developing custom functionality for your store.

> **Sign up for email alerts**
>
> WooCommerce is great at e-commerce but they aren't great at email marketing. The only way to get posts mailed directly to you is to leave a comment on one of their blog posts and check the box **Notify me of new posts by email**.
>
> I have email alerts set up for this site and review the new posts every few weeks.

Annual conferences

In addition to the developer blog, the Woo team has participated in WooSesh (https://woosesh.com/) for several years, unveiling their *State of the Woo* presentation. WooSesh is a free online conference generally lasting 1-3 days across several time zones.

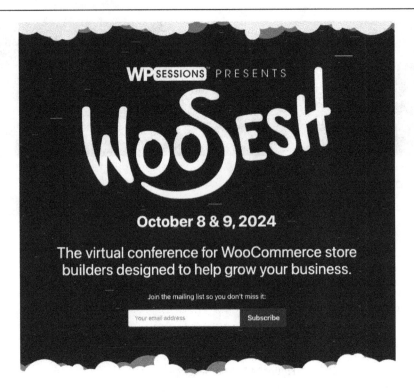

Figure 14.6: WooSesh

I actually helped organize the first few WooSesh events with my co-founder, Brian Richards. He still runs the event and he runs it incredibly well. They gather really smart people doing inspiring things in the WooCommerce world and they share that information.

The *State of the Woo* presentation is generally two hours long, and it covers what the WooCommerce team did in the last year and what they are working on releasing in the next year. This is a really condensed and visually engaging way to understand what's changed and what's coming next.

If you want to take another step on the journey to mastering WooCommerce, I'd pencil in the next dates for WooSesh.

Contributing to WooCommerce

If you are a developer or you have a solid understanding of the code you can have a much more impactful relationship with WooCommerce. You can do two important things:

- You can report issues to the WooCommerce team
- You can submit pull requests to solve issues for the WooCommerce team

Let's look at each of these in more detail.

The repository for the WooCommerce code itself is located at `https://github.com/woocommerce/woocommerce/`.

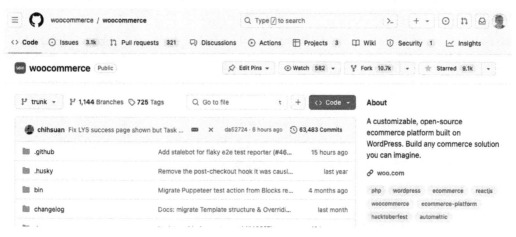

Figure 14.7: Repository for the WooCommerce code

You can immediately see this is an active code repository. There are over 3,000 open issues and over 319 open pull requests solving issues.

If you think you've noticed a bug in WooCommerce, the first thing you should do is search through the open and closed issues. There are just shy of 24,000 closed issues, so many issues have already been discussed within the Woo team. It's very likely someone already filed a similar issue and it's been discussed and planned.

If you've looked through the open and closed issues and don't see your issue mentioned, then open a new issue.

You can type whatever you want into your issue. But if you want your issue to be noticed quickly and get resolved in a reasonable timeline, then you should do a few things:

- Be straightforward and direct. Tell the developers exactly what is not working as expected.

- Make sure the issue works with a fresh installation of WooCommerce and Storefront. If you don't rule out other themes and plugins, many issues are actually the fault of third parties and the WooCommerce team can't invest the time to confirm whether your issue is a WooCommerce issue or a conflict between WooCommerce and another system.

- Include a screenshot or, better yet, a short video showing the problem. Paragraphs of text can be hard to parse, even when well written. A graphic or a video can often show the problem in less time than it takes to read a few paragraphs.

> **You can fix conflicts**
>
> You can fix conflicts with third-party plugins and themes. If you notice WooCommerce and a specific third party cause an issue you can document that and report it to both parties. Typically, one party will provide a fix.

If you want to see a real-life example of this process, take a look at this issue: `https://github.com/woocommerce/woocommerce/issues/31099`.

Figure 14.8: Finding an existing issue

It starts out with me finding an existing issue describing the issue I had. Since there was already an issue describing this problem, I added my own thoughts to the issue, I added a link to official documentation describing how something should work, and I added a video (only 18 seconds) demonstrating the problem with just WooCommerce and Storefront.

Outside of the screenshot, I also included examples from Amazon and Walmart to show what their markup looks like.

With the issue fully described with links to documentation, a video demonstration, and real-world examples, I knew WooCommerce would eventually solve this issue. But I didn't want to wait. So, I decided to solve it myself.

I wrote a pull request (`https://github.com/woocommerce/woocommerce/pull/44413`) where I wrote the code that solved this particular issue. You can see the start of the pull request in the following screenshot:

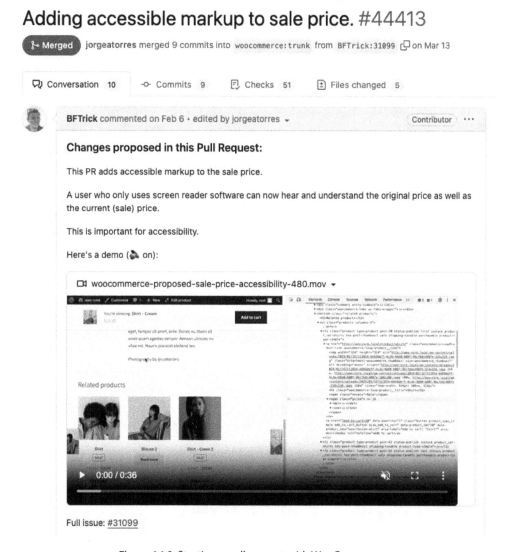

Figure 14.9: Starting a pull request with WooCommerce

You can view the full back-and-forth with the WooCommerce team. They have excellent code quality and hygiene. We iterated on my solution several times making the code defensible and future-proof so no one has to fix it in a few months.

After a few weeks of back-and-forth, this is slated to go out with WooCommerce 8 . 8. So, if you're using that version or a newer version, you're benefiting from me discovering a particular problem on my site and writing the code to fix it for everyone. This is the huge advantage of open source software.

Now that you know the basics of opening issues and submitting pull requests, there is one more aspect of staying up to date with WooCommerce and getting attention on your issues and pull requests, which is knowing what the office hours are.

Office hours

The WooCommerce team holds monthly text-based office hours on Slack. You can find these announced on their developer blog (`https://developer.woocommerce.com/blog/`).

Jacklyn Biggin (Woo Developer Advocate) 10:00 AM
⋯ **\<Begin Monthly Developer Office Hours\>** ⋯
==

Hey **@channel**! 👋 We're here to kick off April's Developer Office Hours! This month, we're joined by @Luigi Teschio, @Thomas Roberts and @Alex Florisca - three developers focused on things like blocks, the site editor and beyond! Team, can you introduce yourself to the channel?

We're here for the next hour to answer any questions you have about building on WooCommerce. Got questions? Let us know! 🐾 (edited)

👣👣👣 10   🐾 2   🙌 1   😊⁺

Figure 14.10: Office hours on Slack

This is the best way I've found to move forward on my issues and pull requests. There are usually a few developers available to answer your questions, and if you come prepared, you can get them to look at several of your issues and help you narrow down whether it really is an issue with WooCommerce, how you might be able to solve it, and how you can write a pull request solving this issue for Woo.

Additionally, these office hours often have a focus where you can discuss new features the Woo team is working on, such as **New Product Editor** in *Figure 14.11*.

I've used office hours to get attention on some of my recent issues, and the extra acceleration meant I was able to get issues solved in weeks instead of months or years.

Summary

Security is important for every business. By configuring a few best practices now and working with the right host, you'll have a pretty secure business. You also know where to start in the wide world of website accessibility. By knowing how to contribute code to WooCommerce, you actually have the agency to affect all of these not just for your site but for all WooCommerce sites.

We're at the end of the book and you are well on your way to mastering WooCommerce. You have the essentials documented in this book.

You can build your own WooCommerce store, sell the products you want to sell, display those products how you want, and customize a theme for your brand, including landing pages and customizing the product page.

You can also understand what's happening in your store by viewing WooCommerce sales data, you can accept payments in person with a POS system, you can configure any number of shipping services to fulfill your packages worldwide, and you also understand the basics of customizing WooCommerce with code and even building your own WooCommerce plugin.

While you have the essentials in this book, the rest you'll learn when you build or improve your WooCommerce store. So, my advice is to deploy one thing you learned in this book this week. The more you iterate and move forward, the better off you are in the world of e-commerce.

Index

Symbols

360-degree images
  adding, to products 259, 260
  displaying 258

A

A/B testing 272
active product filters 83
ad-free experience
  creating 7
  custom plugin, making 9
  Jetpack Without Promotions 8
  marketplace suggestions, disabling 9
Ahrefs
  URL 97
always-on marketing 95
Amazon S3 Storage for WooCommerce
  reference link 53
annual conferences 313, 314
**application programming
    interface (API) 144, 169**
Astra 228-230
  reference link 228
Autoptimize
  setting up 201-203
  URL 201
average order value (AOV) 129

B

basic WooCommerce plugin
  activeness, checking of
      WooCommerce 283-285
  building 282, 283
breadcrumbs 69
  configuring, for search engines
      and users 98-100
breadcrumbs, customizing
  personal home page (PHP) code,
      adding to theme 100, 101
  plugins, using 102
browser caching 212
bulk updater
  using 210-212

C

caching
  configuring, via HTACCESS 213
caching plugins
  configuring 212, 213
Category report 129
child theme
  for extensive customizations 240
  reference link 239

Chrome User Experience
    Reports (CrUX) 194
comma-separated value (CSV) file 136, 137
    content, including in 138
    importing 139-142
    products, importing via 139
comparison guide, payment gateways
    reference link 15
Composite Products extension
    reference link 60
concatenation 204
configurable bundles 58-60
consumer key 145
consumer secret 145
content
    optimizing, above fold 214, 216
conversion rate optimization (CRO) 304
Core Web Vitals scores 193
credit card readers, WooCommerce POS
    reference link 152
CRO experiment 273, 274
CSS resources
    minifying 201
Cumulative Layout Shift (CLS) 194
customer acquisition cost (CAC) 93
customer acquisition strategies 94
    always-on marketing 95
    one-off marketing strategies 94
custom tab
    adding 237, 238
custom tab plugin
    installing 236, 237

D

data
    syncing, manually 162

Developer Blog
    following 312, 313
Digital/Downloadable Product Handling
    reference link 52
digital products 50
    defining 50, 51
    downloadable, not virtual 51
    downloads, accessing 53
    large downloadable files 53
    virtual and downloadable
        products, configuring 51

E

e-commerce, landing page 268
    add-to-cart button, adding 270
    add-to-cart button, customizing 271
    featured product, adding 268-270
    One Page Checkout 272
    product ID, finding 270
e-commerce sites
    limitations 6
email marketing strategies
    reference link 21
end-of-life (EOL) 309
enterprise-level ERPs
    references 143
Enterprise Resource Planning
    (ERP) system 135
    enterprise-level ERPs 143
    entry-level ERPs 143
    finding 142, 143
    integrating with 142
    mid-level ERPs 143
    products, importing into 146, 147
    reference link 148
    using 148

entry-level ERPs
references 143

F

files
concatenating 204, 205
Finale Inventory 144
configuring 144-146
First Input Delay (FID) 194
FOMO
events, removing 250
notification frequency, customizing 252, 253
notifications, customizing 250
notifications, hiding on mobile 251, 252
setting up 242-250

G

Google
XML sitemap, submitting to 104, 105
Google Analytics 132
Google Search Console 105-107
Google Trends
URL 97
gross profit 129
grouped products 54-56
GrowthBook
URL 273
GTmetrix 191, 194
free plan limitations 199
periodic testing, configuring 199-201
starting point, measuring 191, 192
GTmetrix Grade 193

H

heatmaps
setting up 274-276
hidden widgets 82, 83
High-Performance Order Storage (HPOS) 5
hook visualizers
installing 231-233
Hotjar
information, adding for new site 278
installing 276
site, adding to 277
siteID, copying from 278
HTACCESS
caching, configuring via 213

I

images
optimizing 205
optimizing, with Jetpack 205, 206
Imagify 205
bulk updater 210-212
images, optimizing 208-210
URL 208
inheritance 293

J

JavaScript resources
minifying 201
Jetpack 205
images, optimizing 205, 206
site accelerator, enabling 206-208
Jetpack Without Promotions 8

K

key performance indicators (KPIs) 131
keyword research, for e-commerce 95
  list of keywords, creating 96, 97
  process, optimizing for keywords 98
  search volume, comparing 97, 98
Klaviyo
  URL 21

L

landing page
  building 262, 263
  content, adding 265-268
  e-commerce, adding 268
  structure 263-265
Largest Contentful Paint
      (LCP) 193, 194, 230

M

Mailchimp
  URL 21
margin 130
Metorik 131, 132
  URL 131
mid-level ERPs
  references 143
monthly text-based office hours
  on Slack 318
Moz
  URL 97
multiplexing 204

N

new features, WooCommerce 4
  High-Performance Order Storage (HPOS) 5
  WooCommerce Admin 4
  WooCommerce blocks 4
  WooCommerce Payments 4

O

one-off marketing strategies 94
orders
  fulfilling, with Shippo 174-177
  refunding 121
  refund requests 122, 123
orders, fulfilling 112
  boxes, packing 116-118
  New order badge, in site admin 113
  order notifications 112
  orders, browsing 113
  orders, marking as complete 121
  packages, dropping off 120
  shipping information, viewing 114, 115
  shipping labels, printing 118-120
order statuses
  customizing 285

P

packages
  fulfilling, with ShipStation 181-183
page caching 214
PageSpeed Insights
  URL 191
Pantheon
  URL 6
Patch Stack
  URL 309

Payment for Stripe
  reference link 152
  setting up 152-154
payment gateways 15
PayPal Payments
  URL 15
periodic testing
  configuring, in GTmetrix 199-201
personal home page (PHP) code 100
Plug and Play integrations 143
plugin files
  creating 282
Point of Sale (POS) 149
post status
  adding, to WooCommerce 288
  registering 287
Pressable
  URL 7
procurement tools 143
product archive pages
  optimizing 71
  URLs, simplifying 75, 76
product blocks 84-90
  customizing 89, 90
product bundles 54-58
  reference link 29
product categories 68
  collectively exhaustive 69, 70
  context 74
  description 75
  descriptions, writing for 72, 73
  meta-description 74, 75
  mutually exclusive 69, 70
product data tab
  adding 236
product filters
  active product filters 83
  adding, to Shop page 77-82

product kits 60
product page
  extra tab, adding 254-257
  rearranging 230, 231
  social proof (FOMO), adding 241, 242
  video tab, adding 253
product price
  moving 234-236
products
  featuring 90
  importing, into ERP 146, 147
  importing, via CSV 139
  tagging 70, 71
public roadmap, WooCommerce
  reference link 5

R

redirects
  in WordPress 76
refund process
  building 124, 125
REST
  reference link 144
REST API 144

S

sales data
  viewing 125
sanitization functions, WordPress
  reference link 299
scroll maps
  setting up 274-276
search engine optimization
    (SEO) 36, 93, 94, 304
Search Engine Results Page (SERP) 74, 98
selling channel 179

SEMrush

URL 97

**settings page, building with
    WooCommerce 289**

constructor, creating 293-295

field settings, adding 295-299

Integration child class, creating 293

main integration file, creating 290-292

shipping information

custom integration, building 169

daily email, processing 170

data, updating 169, 170

emails, sending 166, 167

order, retrieving through custom
    integration 170

sending 166

updating 166

webhooks, configuring 167-169

Shippo

configuring 171

orders, fulfilling with 174-177

setup information, configuring 173

signing up for 171-173

ShipStation

configuring 177

integrating with 177-181

packages, fulfilling with 181-183

pick lists, printing 184, 185

ShipStation app

using 185, 186

simple products 29, 30

data fields 30

description 37, 38

fully configured simple product 38-40

images, adding 35, 36

inventory 32

optional fields 31

SEO tip, for image filenames 36, 37

shipping 33, 34

short description 38

stock 32

taxonomies 34, 35

Simply Show Hooks plugin 233

single database systems 161

single product pages 89

singular source of truth (SSOT)

reference link 135

site accelerator

enabling, within Jetpack 206-208

Site Kit

reference link 201

social proof 241

Square 154, 161

business location 157

connecting with 154, 155

data, syncing 158

data, syncing manually 159, 160

flat, versus hierarchical categories 160

products, marking to sync 158

selecting 161

setting up, for WooCommerce 156-158

sync settings 157

SSL certificate 309

Stock Keeping Unit (SKU) 158, 184

store

accessible, making 305

color contrast, improving 305, 306

testing, with keyboard 306, 307

Storefront 225

product pagination 227

reference link 73

sticky add-to-cart button 225, 226

Stripe 15

installing 16

URL 15

subscription-compatible gateways

reference link 64

subscription product
  creating 61, 62
subscriptions 60
  failed payments, retrying 65
  manual, versus automatic renewals 64
  recurring payment option, adding
    to product 62, 63
  reference link 29
  settings 63
  switching 65
  synchronization 65
synced databases
  mastering, via API 162

T

tags 68
  adding, to products 70, 71
Taxes report 130
taxonomies 34, 35
  reference link 34
test sites 5
  files, migrating 6, 7
  functionality, testing with publicly
    accessible URL 7
theme
  selecting, for WooCommerce 219
third-party logistics partner (3PL) 165
Total Blocking Time (TBT) 194
traffic generation 93
Twenty Twenty-Four theme
  exploring 220-223
  reference link 220
two-factor authentication 310

U

United States Postal Service (USPS) 116
user experience (UX) 79

V

variable products 40, 41
  attributes 40-43
  images for variation 40, 48
  individual variations, editing 45-47
  multi-attribute products 40
  multi-attribute variations 48
  reference link 40
  variations 40-45
  variations, troubleshooting 40-50
video tab
  installing 254

W

W3 Total Cache
  URL 201
Waterfall 197
web accessibility 302
  business benefits 304
  inclusivity 302, 303
  legal liability 303, 304
  preparations, for demographic trends 305
webhooks 167
website development process 5
website host 308
Web Vitals 193
  URL 193
widgets 82
WooCommerce 3
  Add tax rates 17-19
  contributing to 314-318
  data, exporting out of 136
  installing 10, 11
  new features 4, 5
  payment gateways 15
  payment settings 14, 15

recommended add-ons 20-23
Square, setting up for 156-158
store address and store details, setting 12, 13
store, personalizing 24-26
WooCommerce 360° Image
installing 258
reference link 258
WooCommerce Admin 4
WooCommerce analytics
using 126, 127
WooCommerce blocks 4
WooCommerce Block Themes
reference link 89
WooCommerce Code Reference 233, 234
WooCommerce Cost of Goods
URL 129
WooCommerce example plugin
using 286
WooCommerce Payments 4, 15
URL 15
WooCommerce PDF Watermark
reference link 52
WooCommerce POS 150, 161
credit cards, accepting 152
selecting 154
setting up 150-152
testing 151
URL 150
WooCommerce REST API
reference link 144
WooCommerce, safety and security 308
hosting, checking before launch 308
scan for downtime 312
SSL certificate, using 309
two-factor login, for administrators 310, 311
version of PHP, keeping supported 309, 310
WordPress core and plugins,
    keeping up to date 309

WooCommerce Smart Refunder
URL 124
WordPress
URL 8
WP Engine
URL 6
WP Hooks Finder 231
WP Rocket
URL 201

X

XML sitemap 102
submitting, to Google 104, 105

Y

Yoast SEO plugin
URL 74

Z

Zapier
URL 136

packtpub.com

Subscribe to our online digital library for full access to over 7,000 books and videos, as well as industry leading tools to help you plan your personal development and advance your career. For more information, please visit our website.

Why subscribe?

- Spend less time learning and more time coding with practical eBooks and Videos from over 4,000 industry professionals

- Improve your learning with Skill Plans built especially for you

- Get a free eBook or video every month

- Fully searchable for easy access to vital information

- Copy and paste, print, and bookmark content

Did you know that Packt offers eBook versions of every book published, with PDF and ePub files available? You can upgrade to the eBook version at packtpub.com and as a print book customer, you are entitled to a discount on the eBook copy. Get in touch with us at customercare@packtpub.com for more details.

At www.packtpub.com, you can also read a collection of free technical articles, sign up for a range of free newsletters, and receive exclusive discounts and offers on Packt books and eBooks.

Other Books You May Enjoy

If you enjoyed this book, you may be interested in these other books by Packt:

Joomla! 4 Masterclass

Luca Marzo

ISBN: 978-1-80323-897-5

- Build your websites using Joomla 4's enhanced features
- Explore advanced content-handling features like scheduled publishing options, custom fields, and the workflow feature
- Discover the search engine optimization features included in Joomla 4
- Set up your website to handle multiple languages and structure the navigation system
- Understand the customization features provided by Joomla -- templates, overrides, and child templates
- Find out how to use CLI to operate without accessing the CMS backend
- Design tailor-made graphics by customizing Joomla templates

WordPress Plugin Development Cookbook

Yannick Lefebvre

ISBN: 978-1-80181-077-7

- Discover action and filter hooks, which form the basis of plugin creation
- Explore the creation of administration pages and add new content management sections through custom post types and custom fields
- Add new components to the block editor library
- Fetch, cache, and regularly update data from external sources
- Bring in external data sources to enhance your content
- Make your pages dynamic by using JavaScript, jQuery, and AJAX and adding new widgets to the platform
- Add support for plugin translation and distributing your work to the WordPress community

Packt is searching for authors like you

If you're interested in becoming an author for Packt, please visit `authors.packtpub.com` and apply today. We have worked with thousands of developers and tech professionals, just like you, to help them share their insight with the global tech community. You can make a general application, apply for a specific hot topic that we are recruiting an author for, or submit your own idea.

Share Your Thoughts

Now you've finished *Mastering WooCommerce*, we'd love to hear your thoughts! Scan the QR code below to go straight to the Amazon review page for this book and share your feedback or leave a review on the site that you purchased it from.

`https://packt.link/r/1835085288`

Your review is important to us and the tech community and will help us make sure we're delivering excellent quality content.

Download a free PDF copy of this book

Thanks for purchasing this book!

Do you like to read on the go but are unable to carry your print books everywhere?

Is your eBook purchase not compatible with the device of your choice?

Don't worry, now with every Packt book you get a DRM-free PDF version of that book at no cost.

Read anywhere, any place, on any device. Search, copy, and paste code from your favorite technical books directly into your application.

The perks don't stop there, you can get exclusive access to discounts, newsletters, and great free content in your inbox daily

Follow these simple steps to get the benefits:

1. Scan the QR code or visit the link below

https://packt.link/free-ebook/9781835085288

2. Submit your proof of purchase

3. That's it! We'll send your free PDF and other benefits to your email directly